THE BIG BOOK OF

THE HUMAN BODY

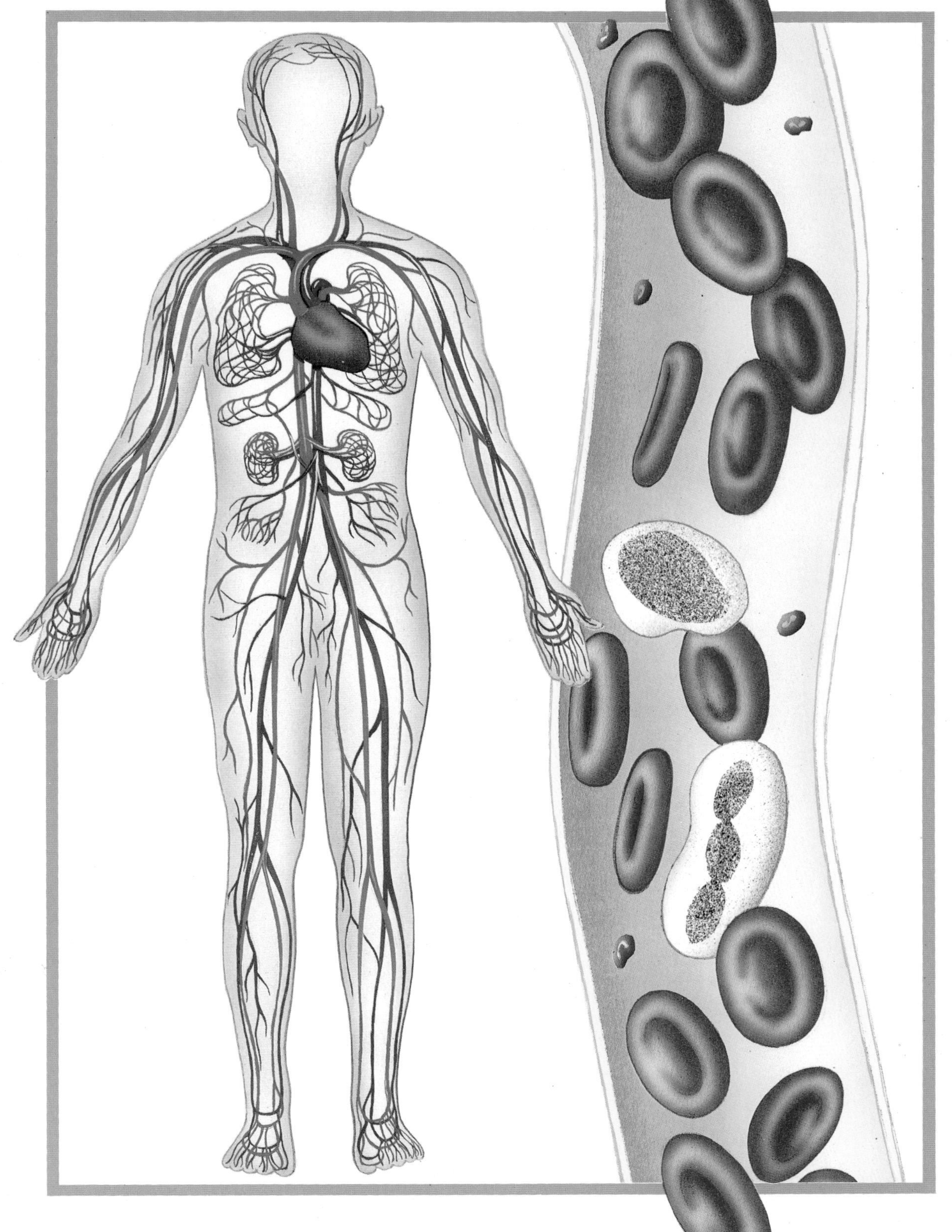

JOHN HARD

This edition published in 1991 by SMITHMARK Publishers Inc.,
112 Madison Avenue, New York, NY 10016
By arrangement with Reed International Books
Michelin House, 81 Fulham Road, London SW3 6RB

ISBN 0-8317-0857-3

Printed in Italy

CONTENTS

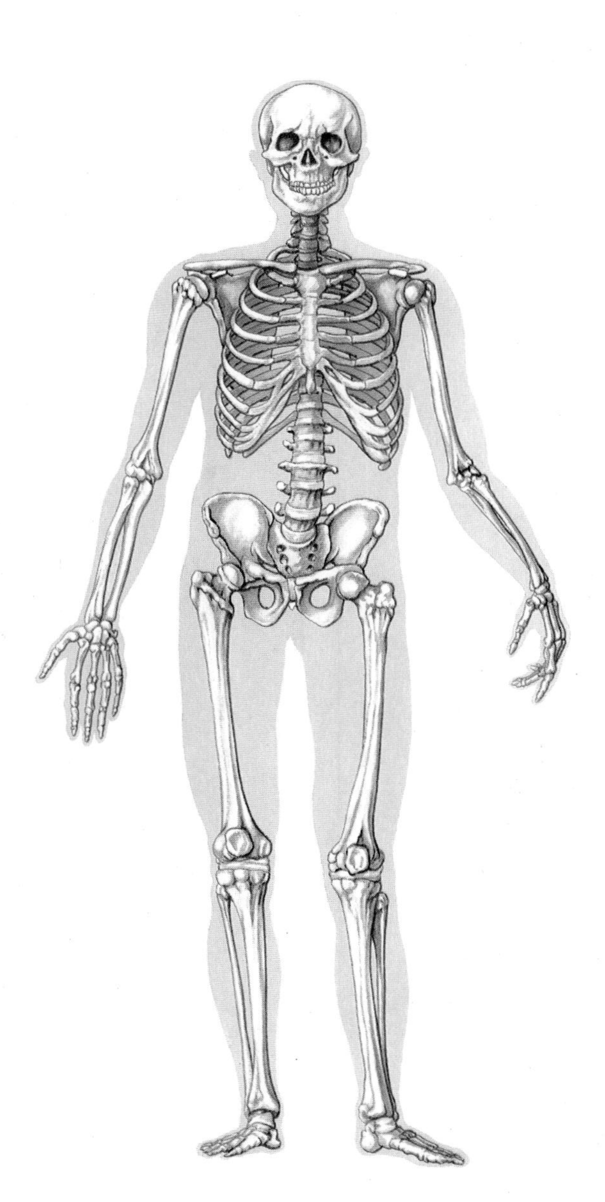

THE HUMAN BODY

Our bodies are marvelously complicated machines which we take for granted for most of our lives. Many people seem to think about their bicycles or cars more than their own bodies, and yet we are so much more complex. It is only when something goes wrong that we ever think about what part of us is doing or how it does it.

Unlike a car parked in a garage, the human body can never stop working, even when at rest. The sleeping body must still breathe, its heart must pump blood, and an even temperature must be maintained. These various systems continue without our thinking about them because part of the brain is constantly active too, receiving and transmitting information. In the pages that follow we shall look at the systems of the body, their structure, how they work, and in some cases the problems that can arise.

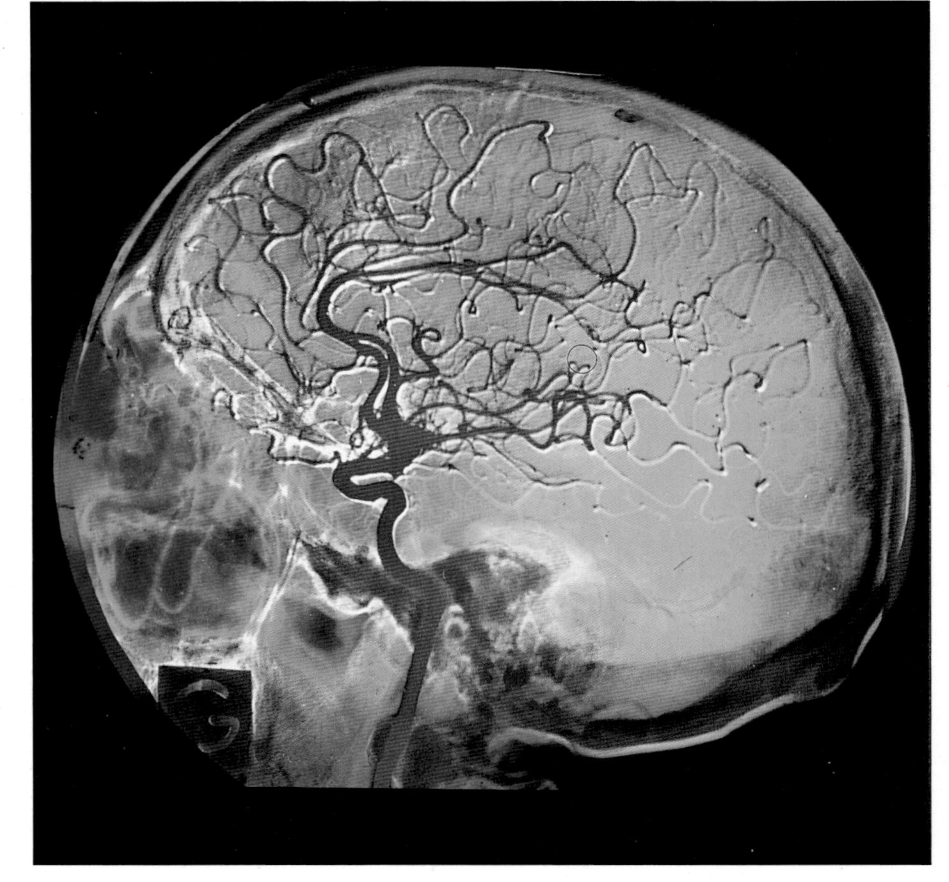

Above right: The human brain is constantly active, directing everything which happens inside the body. In this false-color photograph, the vast network of arteries which supplies blood to the brain and enables it to work is shown in bright pink.

Right: Through play, we gradually learn what our bodies are able to do. Because they have little understanding of danger, young children can be reckless. As we grow older, experience teaches us what the body's limits are.

Above and right: Exercise through play and organized sport has an important role in the development of a healthy body. It helps the muscles, lungs and heart to grow strong. Today children have a wide variety of different sports to choose from and are able to enjoy themselves as they keep fit.

It is important to remember that although the different organs of the body look very different and do completely different jobs, they are all built from cells (see pages 10-11). All the various jobs carried out by cells are controlled by their nuclei and, as we all started as a single cell, all our organs have cells with the same information stored within their nuclei. It is part of the remarkable organization of the human body, and that of all other creatures, that all these different jobs can be performed by the right cell at the right time.

Our bodies also respond to the environment around us. That means the people with whom we live and work, and the physical conditions in which we find ourselves. Toward the end of this book is a section explaining some of the effects that the world around us can have on our bodies and the way that we live.

Finally, although our bodies work well if fed and watered, we can improve or harm ourselves by the way we eat and exercise. Remember that many of the diseases we suffer in the developed countries are partly caused by what we do to ourselves.

CELLS

If you look around you at the enormous range of different plants and animals in the world it is difficult to believe that they are all made up of small units of jelly! But that is really what a cell is: a jelly contained in a very thin skin, or membrane, and made mostly of water, plus some sugars and proteins. Apart from the thin membrane, the other two main parts of a cell are the nucleus and the cytoplasm.

The nucleus of a cell is an opaque solid sphere, containing small strips of special chemicals called nucleic acids, and these carry all the instructions for cell activity in a special code. The nucleus is, therefore, the control center for cell chemistry; it directs the cell in whatever job it is doing. All living plants and animals use the same code to control their chemistry, although some of the jobs carried out by specialized human cells, such as red blood cells, are very different from anything that happens in a plant.

The cytoplasm is the semi-solid jelly that makes up most of the cell, and this is where nearly all of the cell chemistry takes place. Within the cytoplasm are many tiny structures called organelles; these are the sites of the various different chemical processes. The mitochondria are one type of organelle. They are responsible for liberating energy from our food, so any part of the body that is very active, such as muscle or the liver, must have lots of mitochondria in its cells.

The very thin cell membrane controls what enters and leaves the cell. Therefore its most important job is to regulate the delivery of the materials from which the cell will build complicated compounds.

All animal cells are made of these three structures, nucleus, cytoplasm, and membrane, but human cells are further specialized

Below: A greatly enlarged human cell. Strength is provided by microtubules, while chemical production is regulated by membranes and ribosomes.

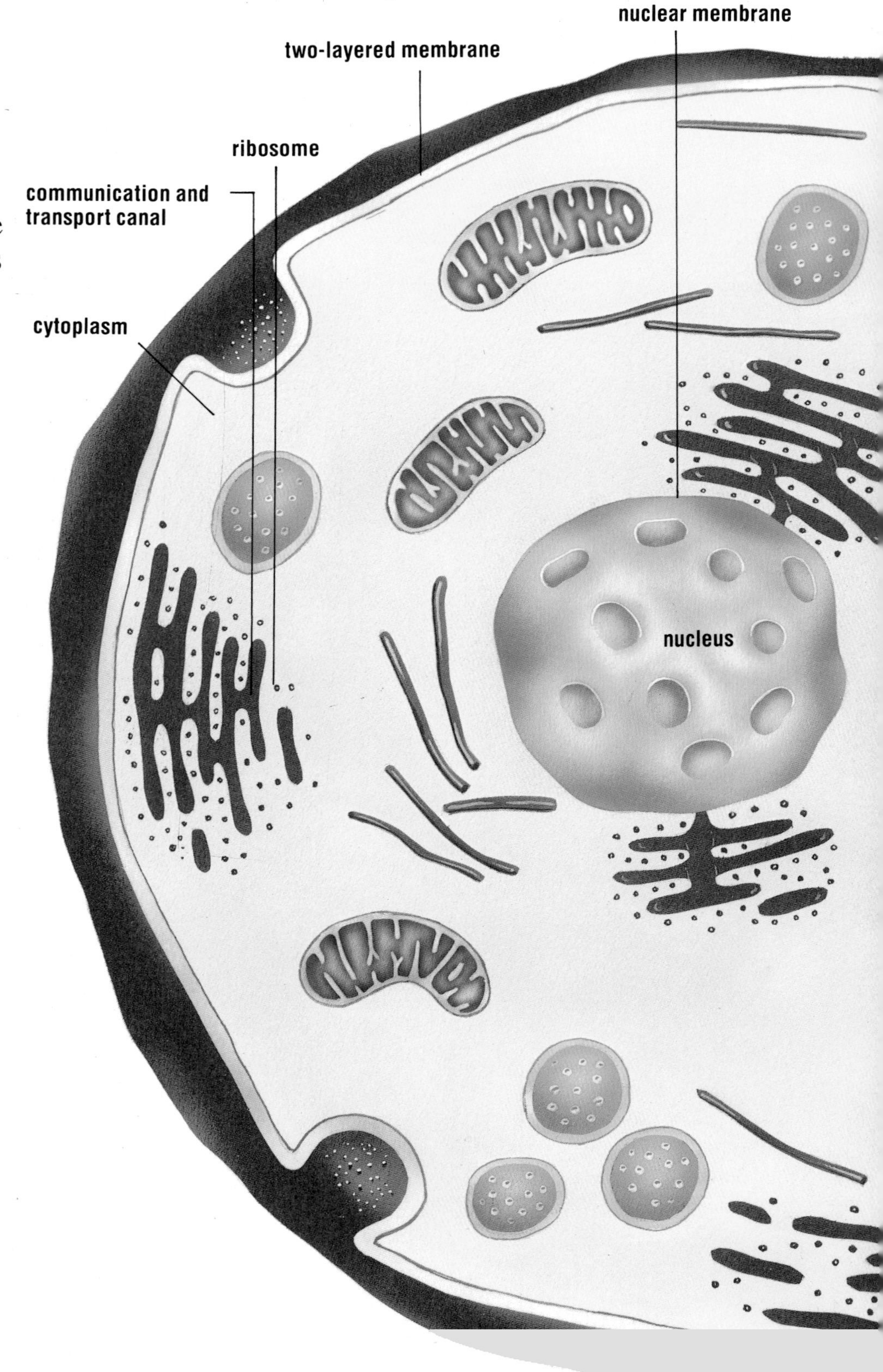

in order to let a cell carry out its particular job. For example, a man's sperm cell has a long tail to help it swim the long journey to join the egg in a woman's Fallopian tubes. Its head contains all the nucleic acid code-messages designed to pass on the man's characteristics to his children. Brain cells have extensions of cytoplasm fiber so that they can join up with other brain cells. This lets a message pass quickly from one cell to many others.

Below: Cells in the tubes of a human kidney. These cells are specialized for the control of water levels in the body.

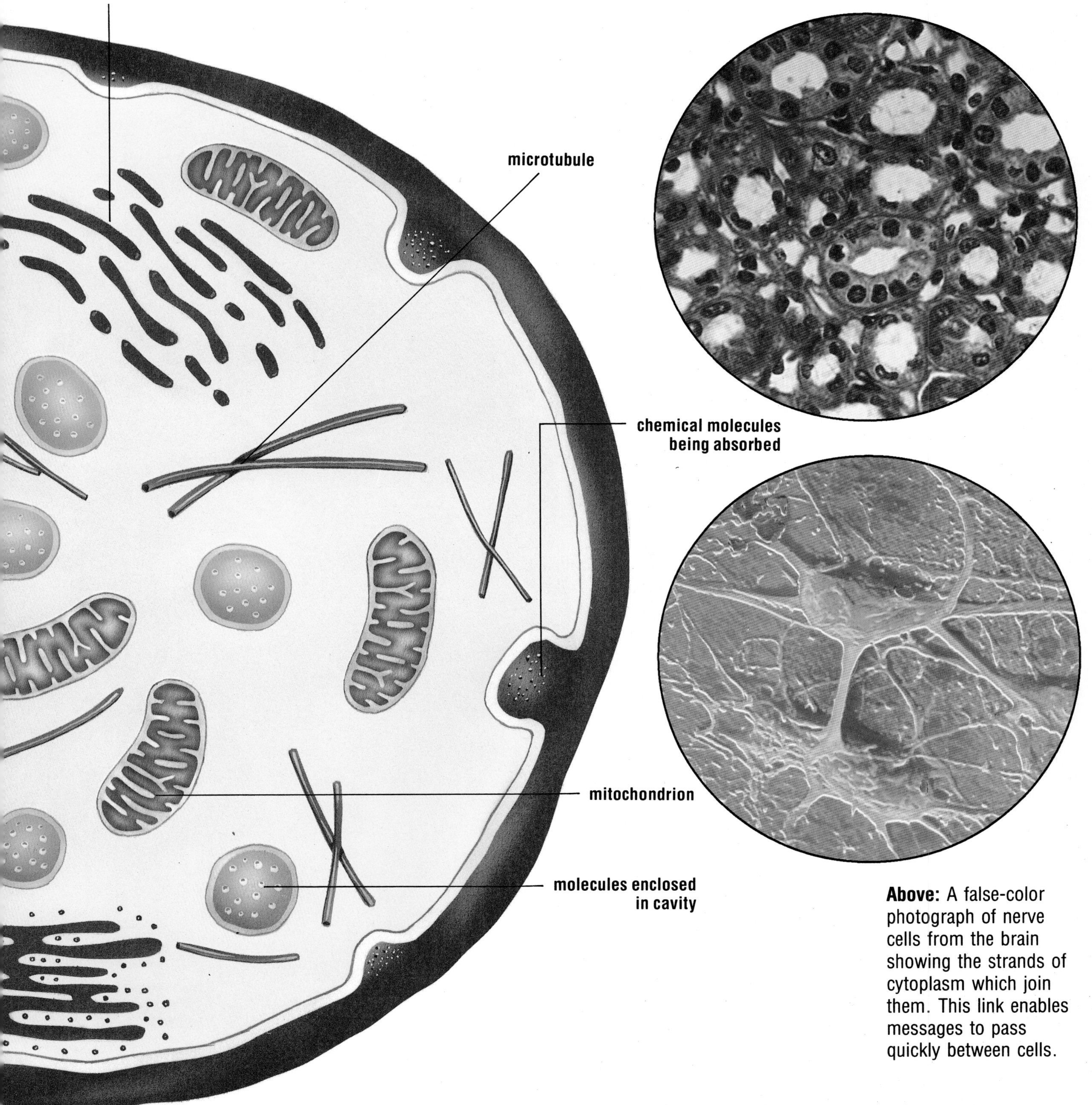

Above: A false-color photograph of nerve cells from the brain showing the strands of cytoplasm which join them. This link enables messages to pass quickly between cells.

SKIN

Most people take their skin for granted — it covers them and acts as a barrier between them and the outside world. In fact, it is the largest organ of the human body and is very active in keeping us at a constant temperature and in our proper shape.

The skin has two main layers. The outer layer is called the epidermis, and is itself made from a single layer of living cells, called the Malpighian layer, and on top of that several layers of dying and dead cells. So, when you look at a person's skin, all you are really seeing is dead cells! Under the epidermis is a thicker layer of tissue, the dermis. This contains blood vessels, nerves, hairs, and muscles to move the hairs.

There are two different types of skin on our bodies. You can see both of them on your arm.

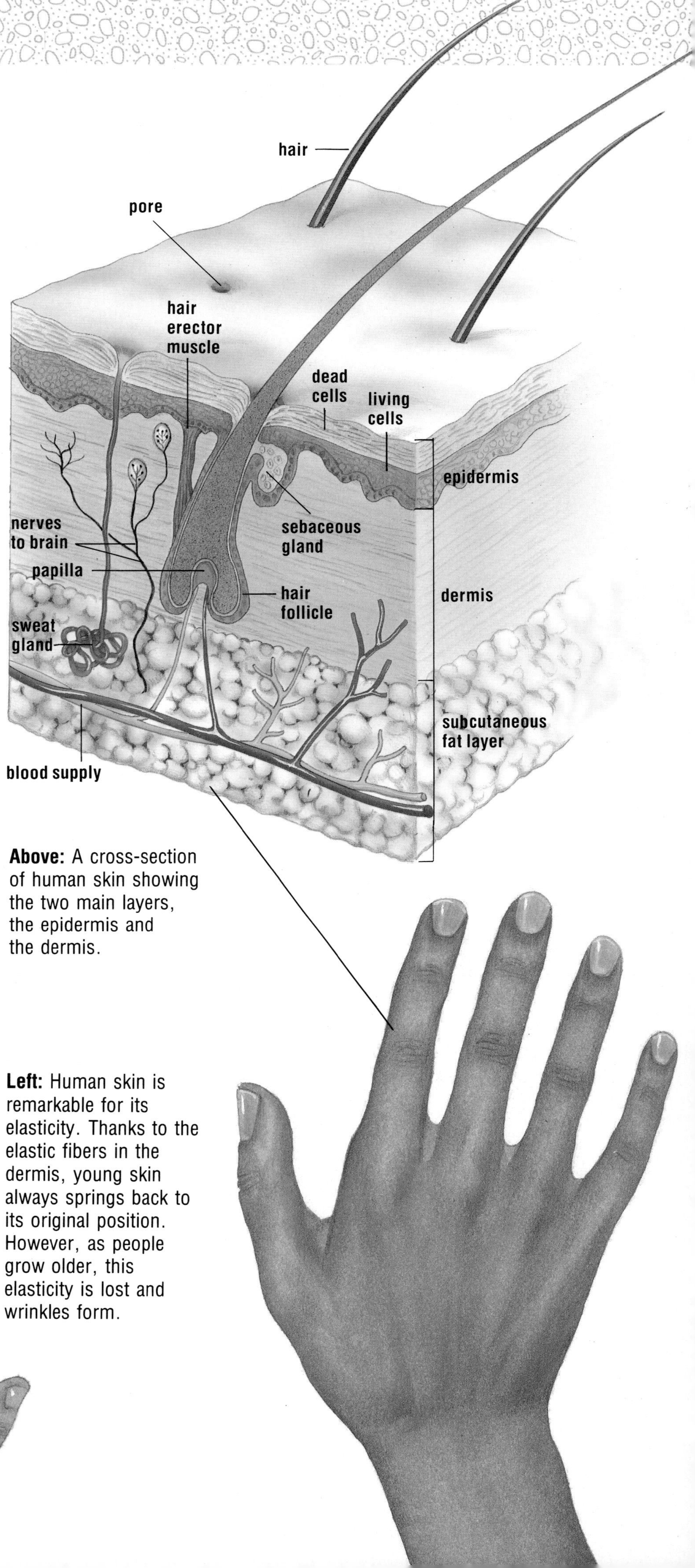

Above: A cross-section of human skin showing the two main layers, the epidermis and the dermis.

Left: Human skin is remarkable for its elasticity. Thanks to the elastic fibers in the dermis, young skin always springs back to its original position. However, as people grow older, this elasticity is lost and wrinkles form.

Above: This greatly magnified photograph shows flakes of dead skin made from old cells falling off the scalp. These cells are continually being shed.

Below: Fingerprints are made from ridges in the epidermis. Everyone has their own unique pattern of whorls.

On the palm of your hand is smooth, hairless skin with lots of protective dead cells — this sort of skin is also found on the soles of your feet. Wherever the skin is often rubbed, thick layers of dead cells build up to form a tough, protective pad.

Now look at the upper part of your forearm. This is the more common sort of skin, the sort that has hairs. If you gently pinch the skin here, it wrinkles and folds easily. This ability to stretch and then return to shape is due to tough white elastic fibers in the dermis that hold the skin in place.

Compare what happens when you push the skin on the upper part of your forearm and then the edge of your heel. Obviously the heel skin is much more rigid because of the thick layers of dead cells, and it cannot move over the tissue underneath nearly so easily.

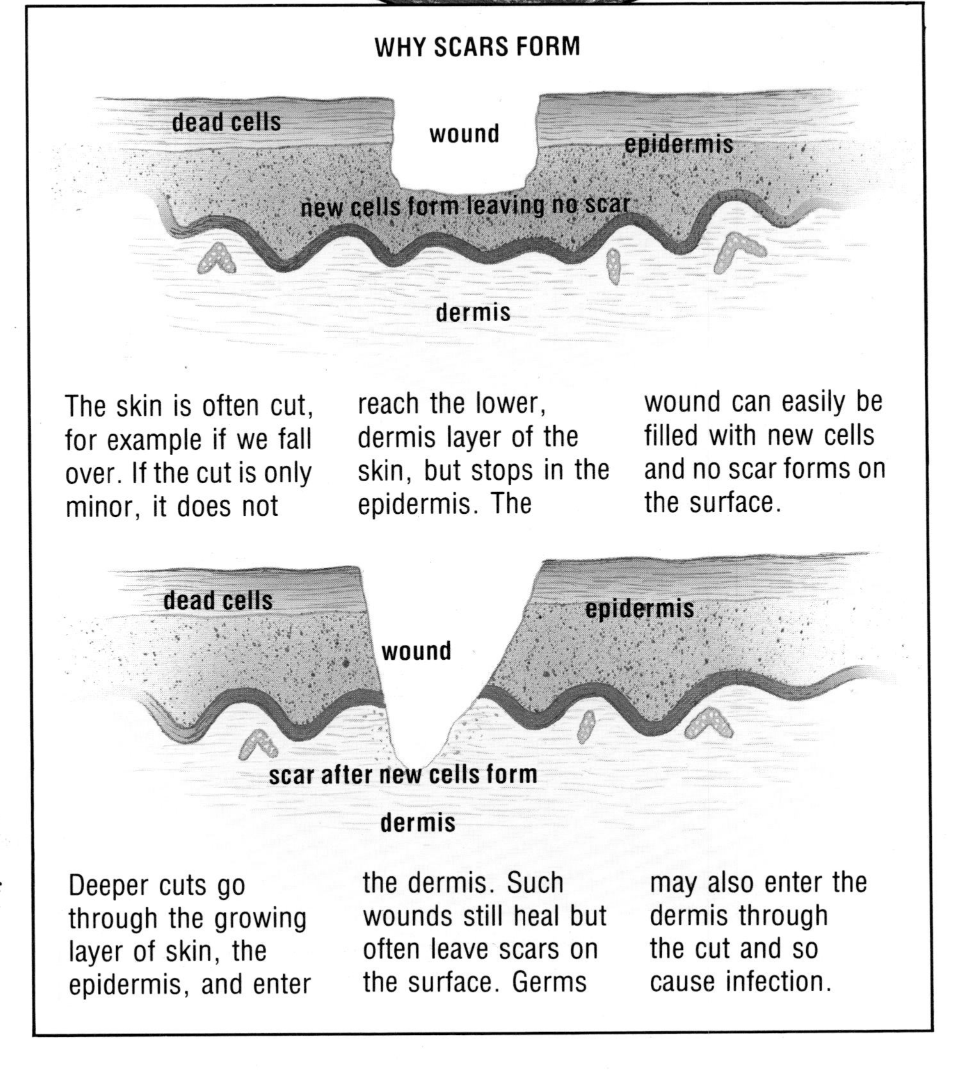

The skin is often cut, for example if we fall over. If the cut is only minor, it does not reach the lower, dermis layer of the skin, but stops in the epidermis. The wound can easily be filled with new cells and no scar forms on the surface.

Deeper cuts go through the growing layer of skin, the epidermis, and enter the dermis. Such wounds still heal but often leave scars on the surface. Germs may also enter the dermis through the cut and so cause infection.

WHAT SKIN DOES

The skin has many jobs to do, and the way and speed that it does them depends upon the messages it receives. All these messages or stimuli are detected by nerve endings. To respond quickly, the nerves must pick up the messages as soon as possible. This is why the sensitive ends of nerves are so close to the surface.

There are different types of nerve endings to pick up the different stimuli, such as pressure or heat. When the ends of the nerves have been stimulated, the message, or impulse, passes to the brain, which recognizes what type of stimulus it is. Some stimuli, for example the rubbing of clothes on the body, keep up a continual flow of impulses, but the brain often ignores such a message if it does not threaten danger.

TEMPERATURE CONTROL

One of the skin's most important jobs is to help keep the body's temperature at a constant level. In many mammals, fur provides an insulating layer of air trapped between the hairs. However, in human beings the "fur" is very thin and on most of the body no longer has any real function. Instead, a layer of fat provides insulation.

A very obvious aid to cooling down is the production of sweat on the skin surface when you are hot. The sweat evaporates, cooling the skin and, therefore, your body,

Far right: We often spend long hours in the sun during the summer. Unfortunately if we stay outside in the heat for too long, our delicate skin tissues may be harmed. Once the skin is damaged, it produces liquid and blisters are formed. Then the skin peels off and nerves are exposed to the air.

SWEATING
Sweat is a salty liquid which is produced in the dermis when the body gets too hot. It travels from the sweat glands to the surface, where it evaporates. This process helps the skin to cool down.

SHIVERING
When the body gets cold, the hairs on the skin stand on end to make a thicker "coat" around us. Then the muscles under our skin move and we shiver. This movement produces heat.

BLUSHING
When a person is embarrassed, blood vessels in their face and neck become wider, or dilate, and fill with much more blood than is usually there. This makes the person "go red" or blush.

down. This is made possible by the small blood vessels just under the skin, which can change their size. When the body is too hot, the blood vessels enlarge, more blood flows along them, and therefore a lot more heat can escape through the surface of the skin. Put your palm near your forehead when you are feeling hot and you can feel that heat escaping. You have probably noticed that some people's faces turn red when they are hot; this is proof that the blood flow has increased.

When you get too cold, the system works in reverse. The blood vessels shrink, reducing the blood flow; with less blood near the surface, heat loss is reduced.

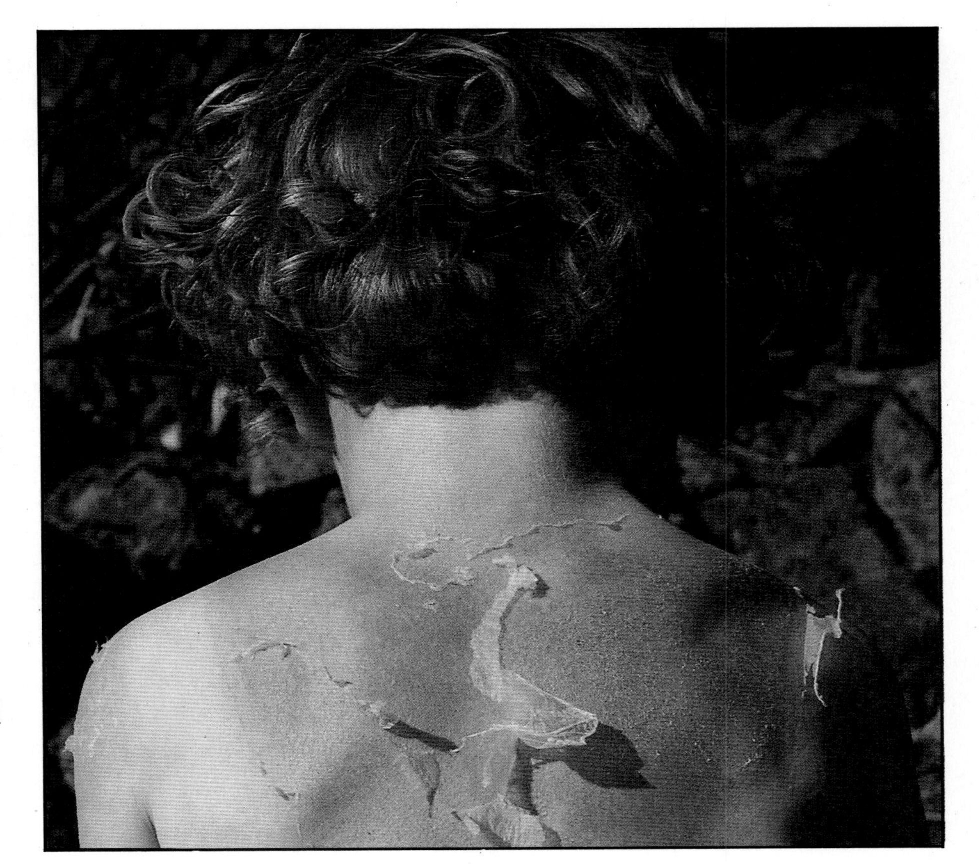

Below and right: Melanin is a pigment which occurs in the epidermis and gives skin its color. It is produced by special cells called melanocytes, which release it into the skin in the form of granules. Black skin contains more melanin than white skin.

melanin granule

melanocyte

Right: One of the jobs of melanin is to protect the skin from the rays of the sun. When white-skinned people sunbathe, their skin produces melanin, which often appears in the form of freckles.

HAIR AND NAILS

Human hair is made of a tough protein called keratin. The growing part of a hair is a layer of cells over a mound at the base of a cavity, called a follicle, deep in the lower, dermis, layer of the skin (see page 12). The mound, called a papilla, contains fine blood vessels which supply the growing cells with food and oxygen. As new cells are formed at the follicle base, the hair grows longer.

The cells on the outside of the hair shaft are very flat and form a protective cuticle. Inside the cuticle is a region of long, tightly packed cells containing colored pigments that give the hair its color. The center of most of the hair that grows on our bodies is hollow, but in the thick hairs of a beard or other adult body hair there are further cells, loosely packed together. The entire hair is kept flexible by an oily secretion from the sebaceous gland which surrounds the upper follicle. Hairs grow rapidly for a while, slow down, and then remain the same length before eventually

Right: Most people cut their nails so that they can use their fingers easily. However it is possible to simply let them keep growing -- in fact they can grow up to 12 in long. This woman's nails have not reached that length yet!

Left and below: The ends of magnified strands of curly, wavy and straight hair show that each one is a different shape. It is partly this shape which causes the three types of hair to fall differently.

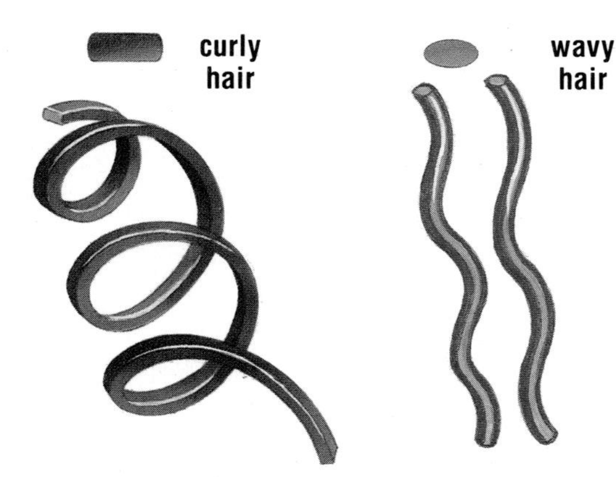

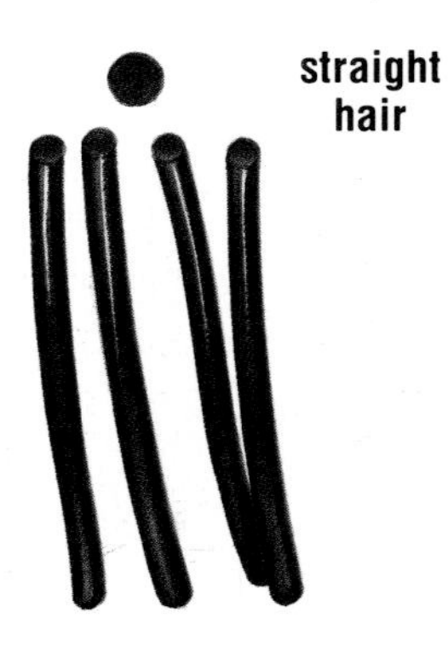

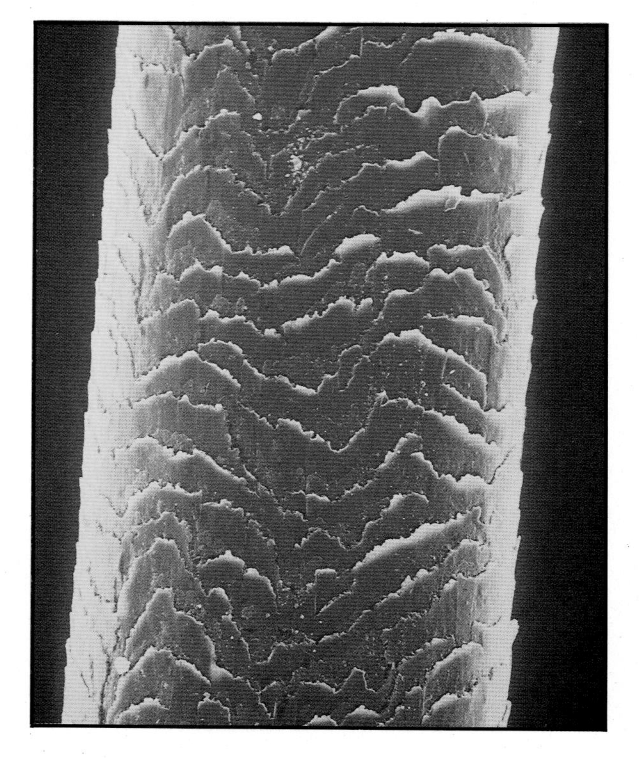

Below: A greatly magnified photograph of a single human hair. It shows clearly the special overlapping cells from which the hair is made. Once hair leaves the follicle where it is formed, it dies. All the cells here are dead, like the cells on the outside of our skin.

dropping out. Later other cells become active and another hair grows from the same follicle.

In mammals other than man the fur is thick and serves various purposes. The most common functions are insulation against extremes of temperature, for example in the polar bear, and protective camouflage against a background, such as in the light brown pelt of the common rabbit. Human beings lack this luxuriant hair. What remains on our heads varies with racial groups and is treated differently depending upon social customs. It has few of the functions found in other mammals.

NAILS

The nails on our fingers and toes are made from the same material as hair, but they form strong flat plates to act as a protective covering. If you have ever lost a fingernail or cut one too short, you will have found out how much they help you manipulate small objects. This is because they steady the tips of your fingers whenever you perform delicate tasks.

Below: Our nails help us to use our fingers in a very accurate way. This is because they provide something for the flesh and bone of the fingers to push against.

Above: Nails grow from the nail bed and protect the ends of our fingers. There is usually about $\frac{1}{6}$ in of nail hidden beneath the surface of the skin.

MUSCLES

Much of the weight of our bodies is due to what is called "flesh." Flesh is made up of muscle tissue arranged in a variety of ways which result in the characteristic shape of the human body.

There are three main types of muscles in the body, and each type has particular features and jobs to do. The most obvious muscles are those attached to the bones, the skeletal muscles. The second most common muscles are those that we cannot control, the involuntary muscles found around blood vessels and the gut. The third, and perhaps most impressive, type of muscle is that making up the heart, the cardiac muscle.

Skeletal muscles are often called voluntary muscles because we can control them when we wish. They are attached to the bones by means of very tough tendons. When a muscle contracts, that is shortens, it pulls on the tendon and this in turn tugs on the bone to produce movement. Muscles can only pull, they cannot expand and push, and this is why they exist in pairs. Each member of a pair works against its partner so that a bone may be moved in two opposite directions. That well-known pair of muscles in

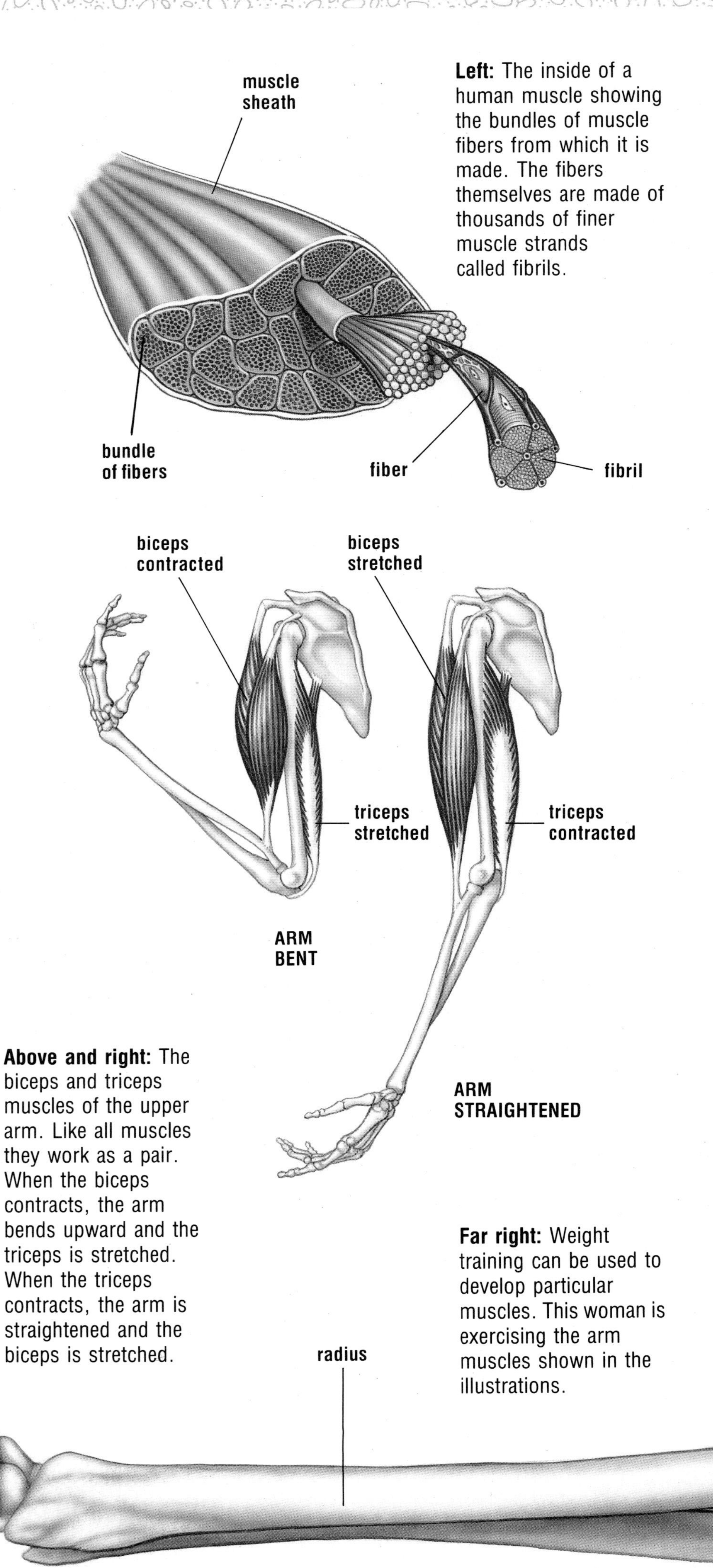

Left: The inside of a human muscle showing the bundles of muscle fibers from which it is made. The fibers themselves are made of thousands of finer muscle strands called fibrils.

Above and right: The biceps and triceps muscles of the upper arm. Like all muscles they work as a pair. When the biceps contracts, the arm bends upward and the triceps is stretched. When the triceps contracts, the arm is straightened and the biceps is stretched.

Far right: Weight training can be used to develop particular muscles. This woman is exercising the arm muscles shown in the illustrations.

the upper arm — the biceps and the triceps — are an excellent example of how your skeletal muscles work. To bend your arm the biceps pulls the lower half of the arm up and stretches the triceps. The triceps does just the opposite when you straighten the arm and stretch the biceps.

It takes a lot of food and oxygen to keep voluntary muscles working, and it is the blood that brings the necessary fuel materials. If a muscle runs out of oxygen while working, it will produce waste chemicals that cause cramp.

Each skeletal muscle has a strong sheath surrounding it. Inside there are bundles of muscle fibers, each up to 1.5 in long. Every fiber is in fact a bundle of many minute fibrils made of linked cells. These voluntary muscles can work very hard and exert enormous power. However, they tire quickly.

The involuntary muscles around the intestines and blood vessels work automatically without our knowing. They contract slowly, but do not tire.

The cardiac or heart muscles are made of tough branching fibers which work continuously throughout our entire lives (see pages 58-59).

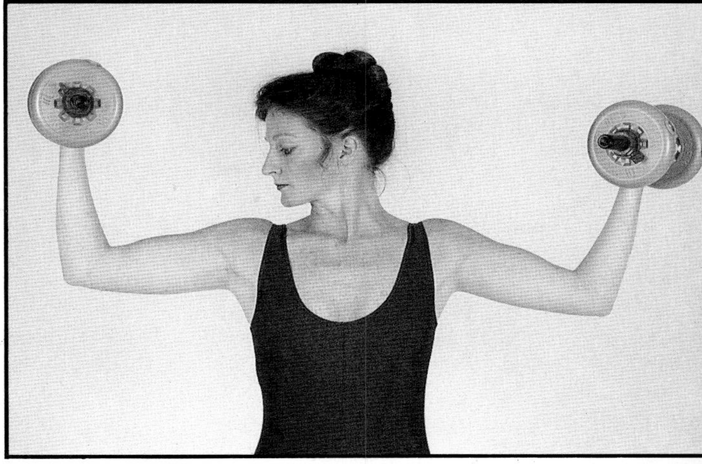

THE SKELETON

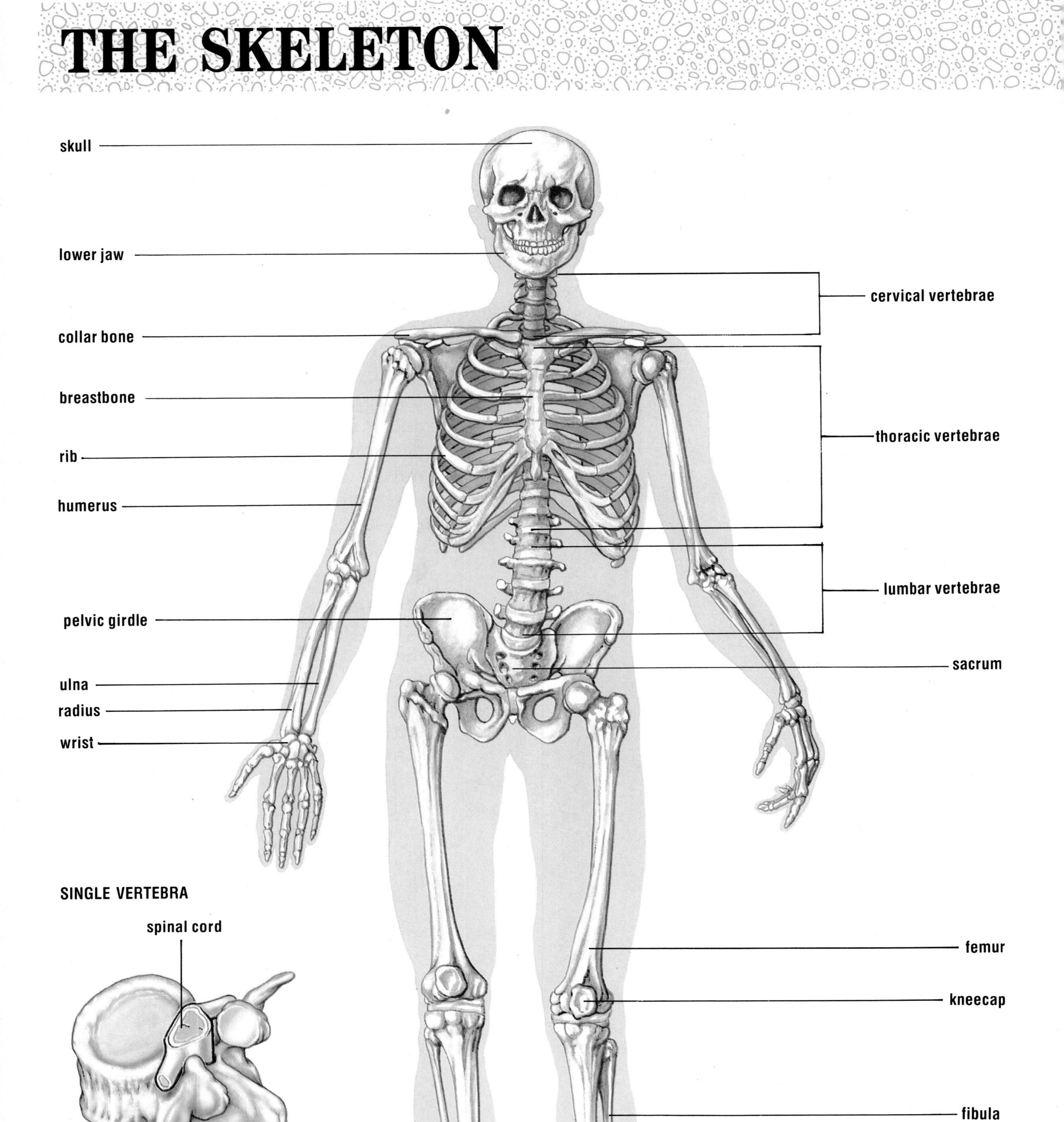

The bony skeleton is the firm framework to which our muscles and organs are attached. It provides a series of supporting structures which have a similar function to the girders of a modern building. If you look at the diagram of a skeleton against the silhouette of a human body, you will see how closely the body's shape fits the form of the skeleton.

The skeleton is divided into two main sections. The skull and spine make up a vertical supporting column of bones, while the limbs, and the horizontal structures they attach to, form the other part.

The skull is only thinly covered in flesh and it is possible to feel the bones, complete with any bony bumps, beneath the skin. The shape of the face is clearly governed by the bones and only the ears and end of the nose can be bent as they are made from gristle. The delicate eyes are sunken into protective sockets.

Far left: The adult human skeleton against the outline of the body we normally see. The skull and the central column of vertebrae provide support for the shoulders and pelvic girdle and the limbs attached to them.

Below left: An adult and a baby both have the same bones in their skull. However the baby's bones do not fuse together until some time after birth. The separate bones can therefore move as the baby is born, making the birth process easier.

Below right: An adult human skull showing clearly where the skull bones have joined.

When you sit in an upright chair you can probably sense both the base of your spine and your pelvis; neither is a flexible bone. The pelvic girdle is what you sit on, which is why you need muscle padding there!

The skeleton has a number of other jobs as well as giving shape to the body. It provides protective boxes for the more delicate organs, such as the lungs and heart within the rib cage and the brain within the skull. Bones also act as levers, upon which muscles can pull so that we can move.

Sometimes the skeleton forms a structure which has more than one job. The pelvic girdle forms a bowl-shaped front to the base of the abdomen, while the fused bones of the sacrum join it at the back to form a bony tunnel. This whole structure protects the organs in the lower abdomen, provides support for the back, and also acts as the point of attachment for our legs.

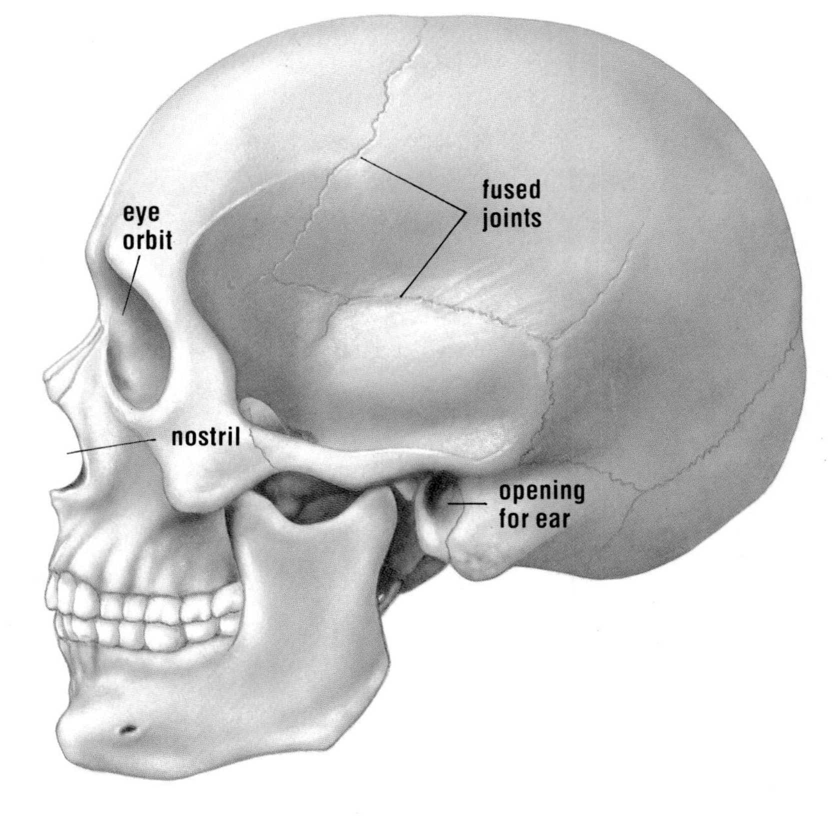

BONES

When we are born we have over 300 bones in our bodies. By the time we have grown up there are about 208. Many of the infant bones have fused, or joined together, to form either large sheets of bone, such as the dome of the cranium (the upper part of the skull), or bigger, more solid blocks, such as the ankle bones.

Some individual bones are very large and their job is obvious. The femur, or thigh bone, is our largest bone and clearly plays a very important part in our major method of moving around. The top end of the femur is shaped like a ball and fits into the socket of the hip to give very free movement. The bottom end is a smooth, curved surface that hinges with the tibia, or shin bone, to form the knee.

The structure of the femur is typical of many of our bones because it is built like a girder. Builders use hollow tubes as scaffolding girders because they can not only bend when placed under stress, but can also spring back into their original shape. A solid rod can be bent, but it is not so elastic. Elasticity is important if our bones are to flex. Bones also contain an elastic compound normally held firmly in place by creamy white bone salts.

The ring of solid bone making up the wall of the femur is filled with a red, jelly-like marrow. This marrow has the job of producing new red blood cells, which explains its color. The marrow in certain other bones, in particular the breastbone, ribs, and spine, produces these cells too. The insides of our bones are also equipped to repair damage, so if

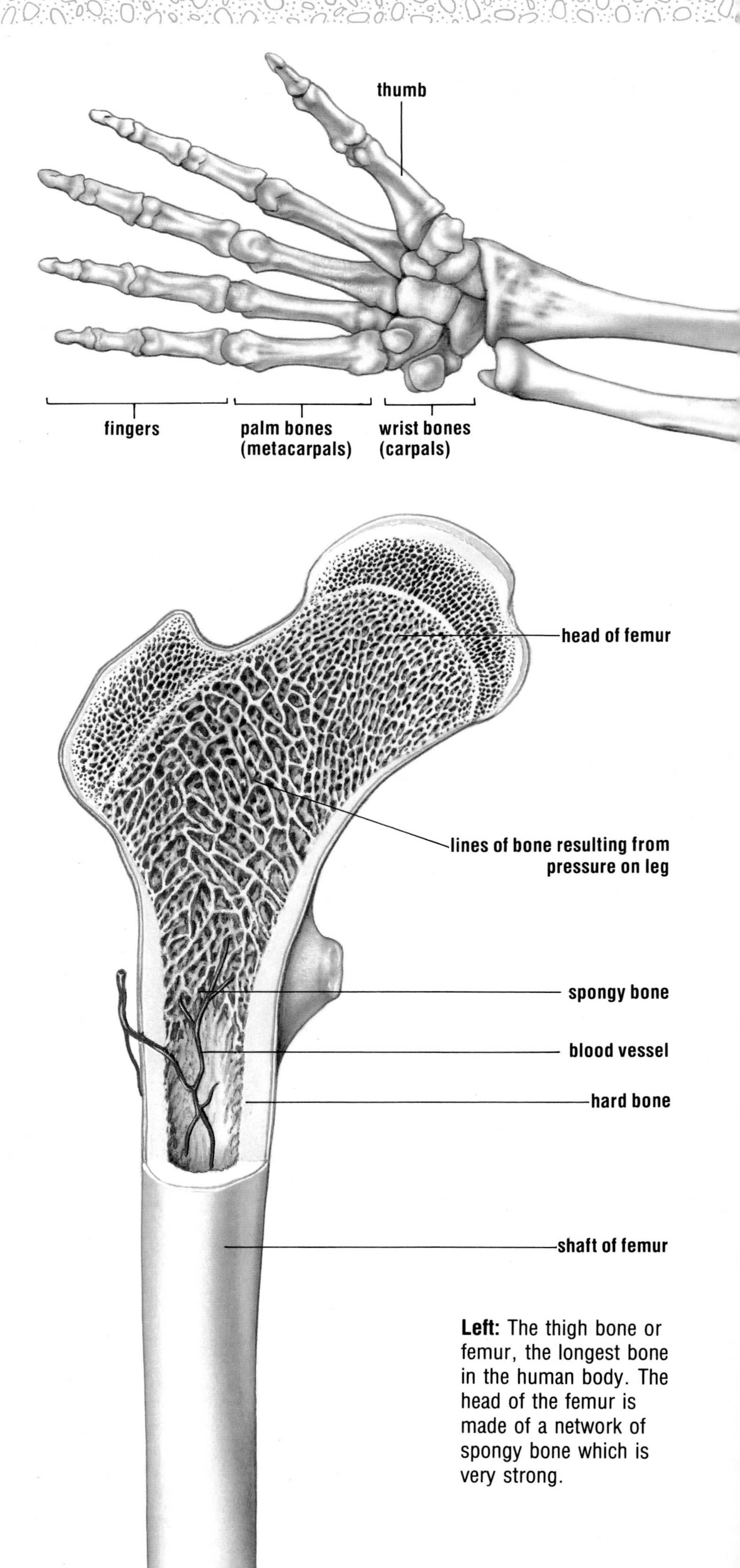

Left: The thigh bone or femur, the longest bone in the human body. The head of the femur is made of a network of spongy bone which is very strong.

Below: The bones of the human arm and hand. The hand alone contains 27 separate bones.

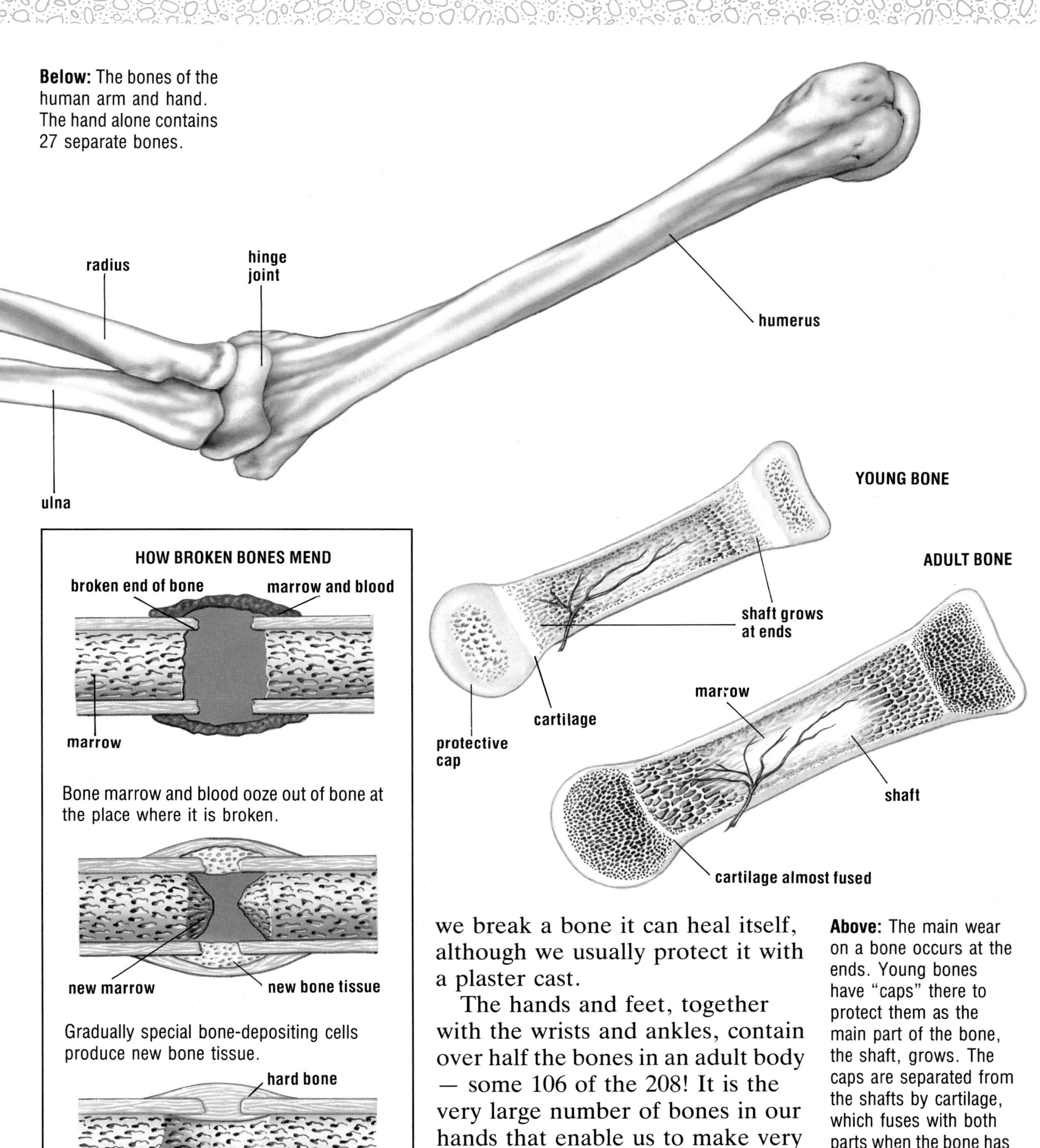

we break a bone it can heal itself, although we usually protect it with a plaster cast.

The hands and feet, together with the wrists and ankles, contain over half the bones in an adult body – some 106 of the 208! It is the very large number of bones in our hands that enable us to make very delicate movements. The fingers are a series of hinged rods and the thumb can bend over the palm to help us grip things. This ability to grip has enabled us to develop and use a wide range of tools.

Above: The main wear on a bone occurs at the ends. Young bones have "caps" there to protect them as the main part of the bone, the shaft, grows. The caps are separated from the shafts by cartilage, which fuses with both parts when the bone has finished growing.

JOINTS

The skeleton forms a strong protective framework around the human body, but to function, this framework has to be mobile. We call the moving parts which let the bones move freely "joints."

There are joints of many types in the body. Ball and socket joints, such as those in the hip and the shoulder, allow considerable freedom of movement. For example, your arm can be made to describe a complete circle in the air. Other joints can only move in a limited way – the elbow joint, for example, works like a simple hinge. You can test this by holding an empty tray in front of you, keeping your upper arms tight against the sides of your chest. Your lower arms will be sticking out to the front, holding the tray. If you want to move the tray to the side, you must twist your upper arms – your elbows can move only up and down, not from side to side. Like elbows, fingers have hinge joints, which can only move in one particular direction.

Another important type of joint is the sliding joint. These joints can be found in both wrists and ankles. Although they can only move in certain directions, they are very flexible and withstand stress well, allowing the feet and hands to respond to the varied types of pressure put on them.

Joints must be able to move without rubbing away the bones of which they are made. They can do this because the bone ends are covered with a smooth type of tissue, called cartilage, and held together by a tough fiber bag which contains a slippery liquid called synovial fluid.

Right: The mobility of the thumb is made possible by a saddle joint, which permits both up-and-down and side-to-side movements.

Below: Ball and socket joints are found in both hips and shoulders. They allow great freedom and ease of movement in many directions.

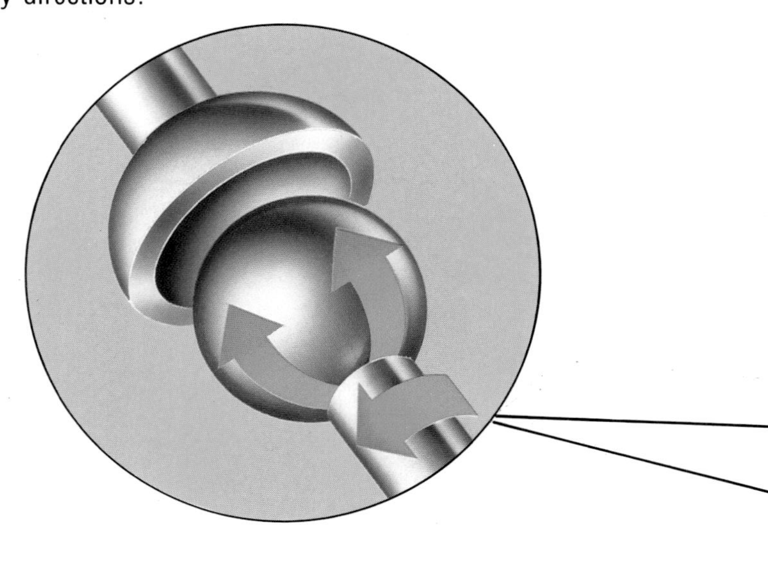

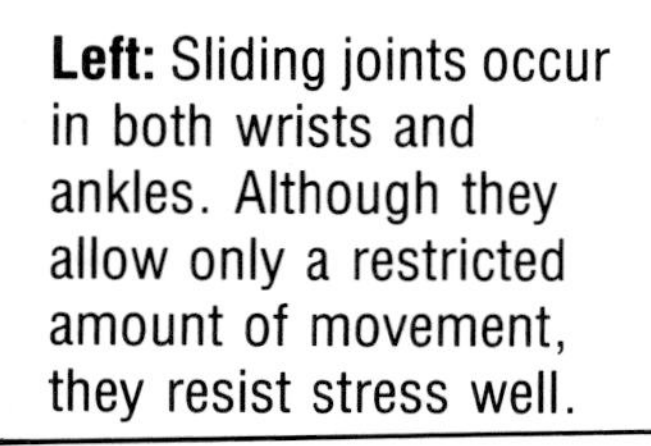

Left: Sliding joints occur in both wrists and ankles. Although they allow only a restricted amount of movement, they resist stress well.

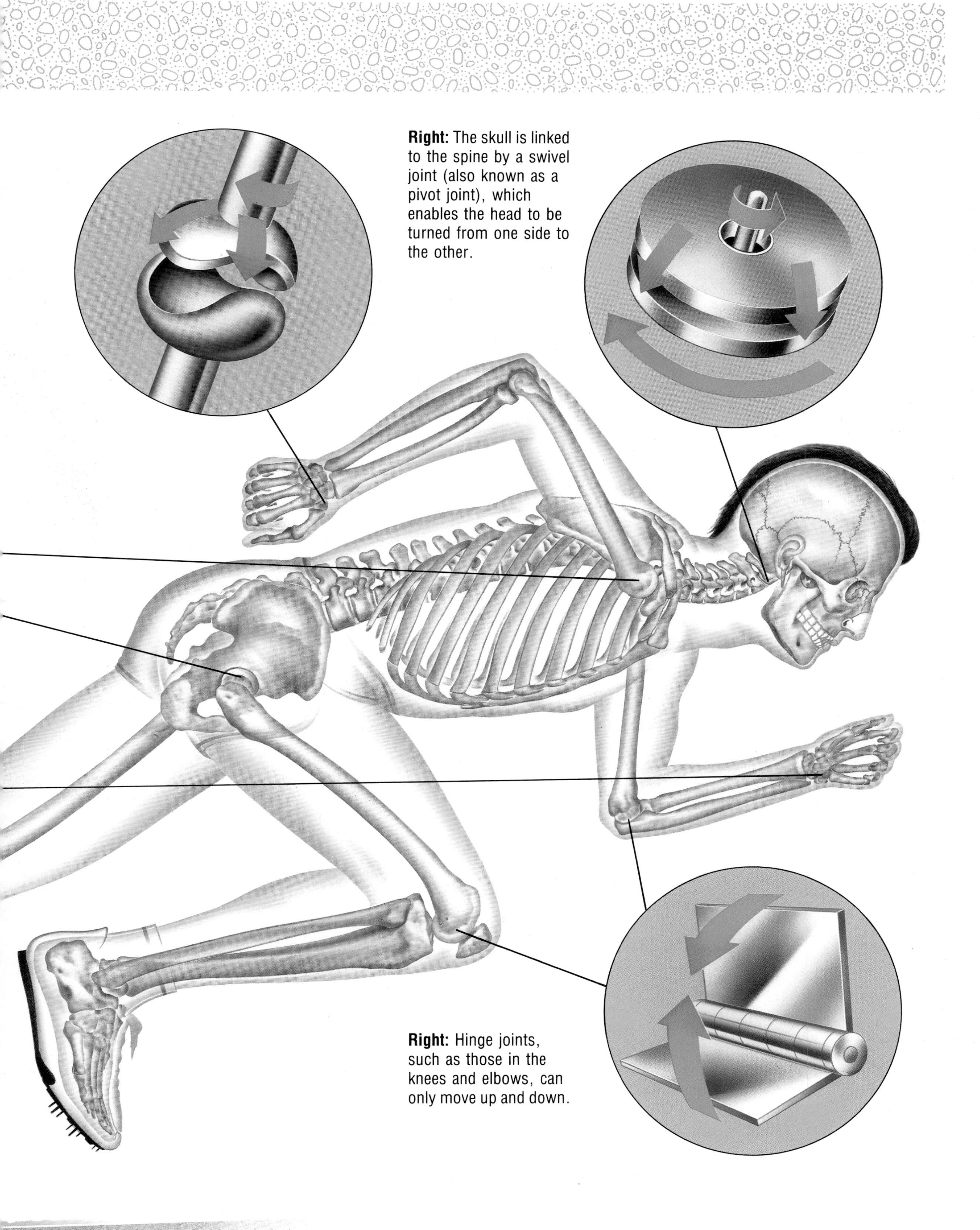

Right: The skull is linked to the spine by a swivel joint (also known as a pivot joint), which enables the head to be turned from one side to the other.

Right: Hinge joints, such as those in the knees and elbows, can only move up and down.

NERVES

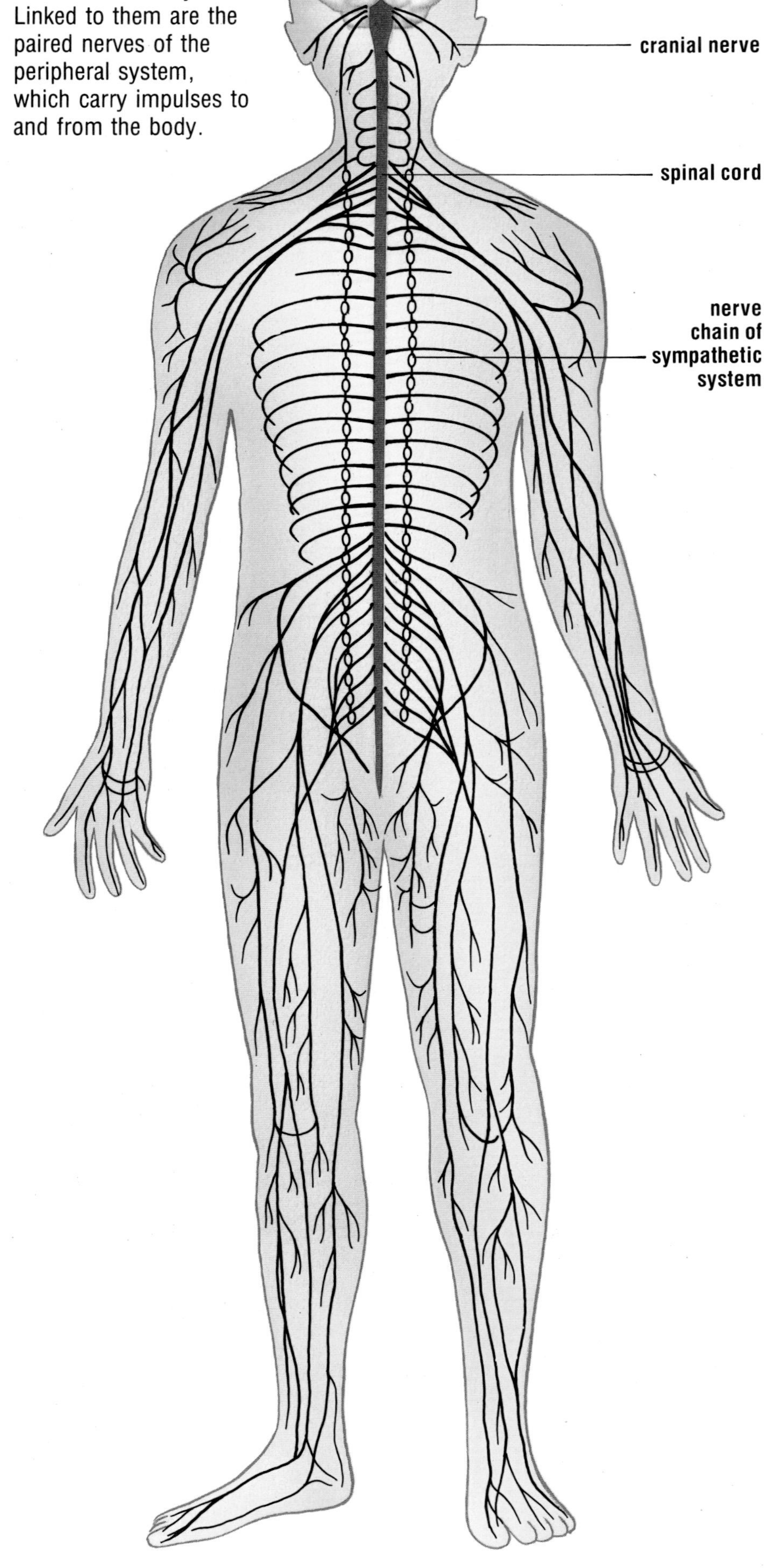

Right: The brain and spinal cord make up our central nervous system. Linked to them are the paired nerves of the peripheral system, which carry impulses to and from the body.

Human beings are extremely specialized animals with very complicated behavior. This means we need great control over our bodies. To achieve such control we have a central control unit which can make decisions and a system for sorting out information and passing on all the decisions the control unit makes.

It is the nervous system which feeds information into the central control unit, the brain, and passes the decisions the brain makes on to the limbs and organs of the body. The brain and the spinal cord (see pages 28-29) make up the central nervous system, and the rest of the nervous system concerned with voluntary activity consists of pairs of nerves running to and from the spinal cord and brain. This is called the peripheral system.

Information reaches the body in many forms — sound, light, scents, touch, and so on — and all these messages are turned into electrical impulses. The impulses are formed at, or just below, the surface of the skin by sensitive cells and nerve endings. Some kinds of information are so specialized that the body has complex sense organs to detect and respond to just these stimuli — the eyes and ears are examples.

Information in the form of impulses travels from the surface of the skin along nerves to the spinal cord. Because the spinal cord is an extension of the brain and structurally part of it, impulses can pass very quickly along its length to the brain. After interpreting the coded messages, the brain decides how to react and instructions pass back along nerves to the part of the body that is to respond.

Some information is fed directly into the brain without passing along the spinal cord. This is the case with impulses carried by the 12 pairs of nerves serving the head and neck and includes messages from specialized sense organs such as the eyes and ears.

Another part of the nervous system, called the autonomic system, is outside our control. Its function is to regulate the unseen chemistry and activities inside us, such as digestion, breathing, heartbeat, excretion, and the production of hormones.

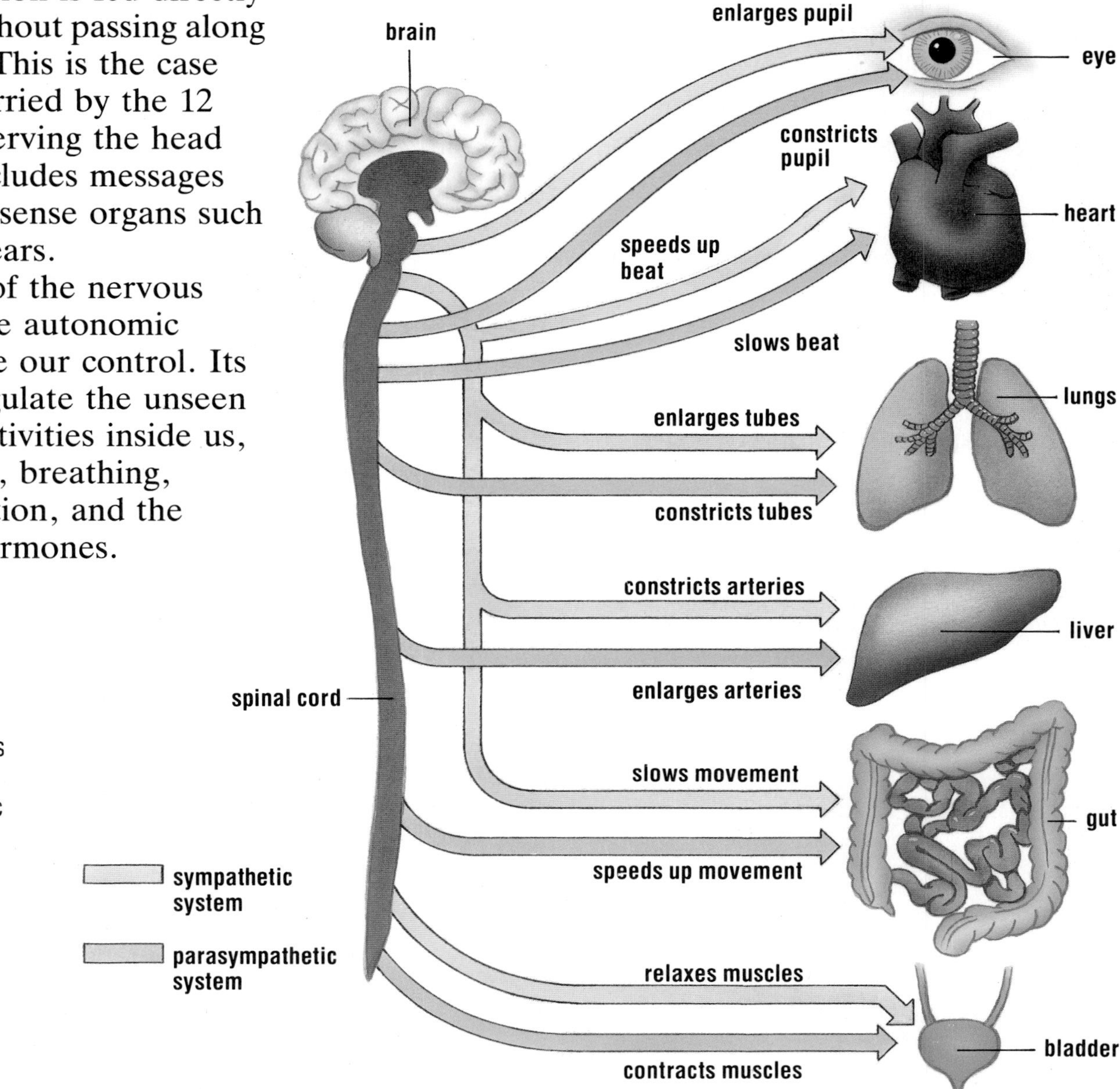

Right: The autonomic nervous system consists of two separate sets of nerves, the sympathetic and parasympathetic nerves. They work in opposing ways on various parts of the human body.

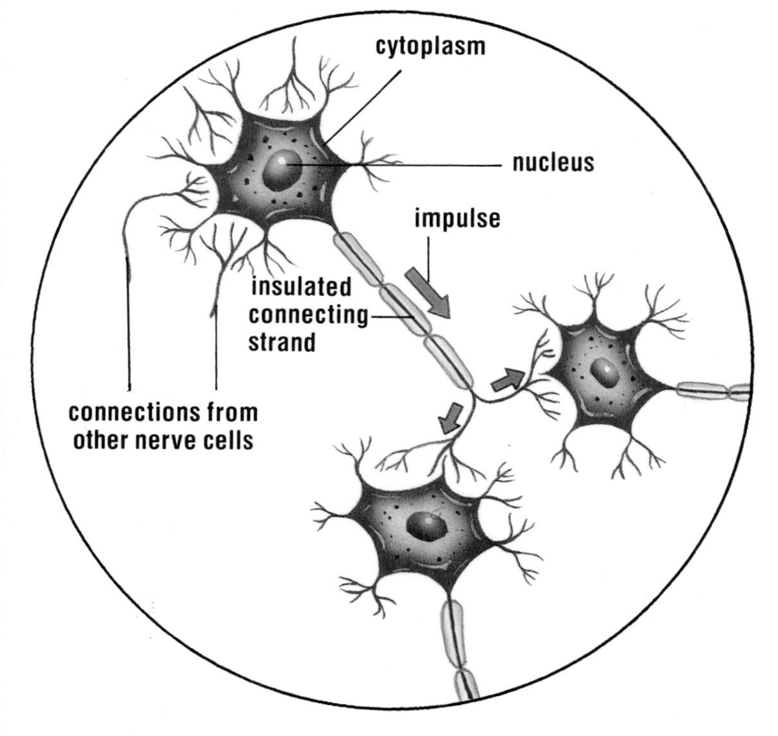

Left: Nerve cells communicate with one another along strands of cytoplasm. One nerve cell may communicate with several others and so send nerve impulses to more than one region.

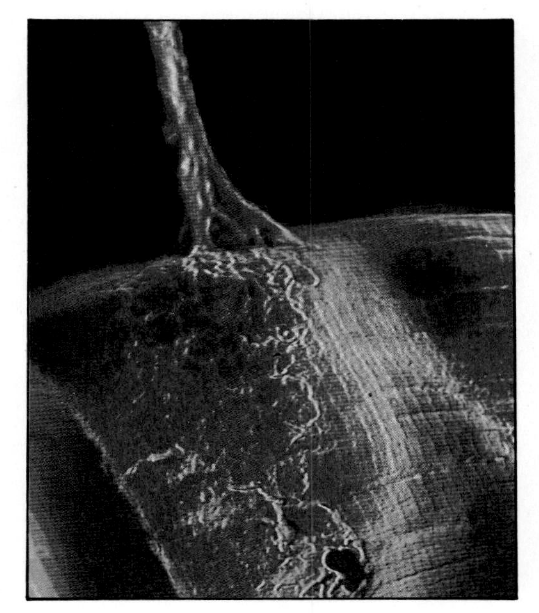

Right: Impulses sent along some nerves cause muscles in the body to contract. This photograph shows a nerve cell linking up with a muscle.

THE BRAIN

At the center of our behavior and control of the body lies the brain. It is a cream-colored structure weighing about 3 lb, with a surface "creased" into many folds. Protected by two layers of tissue called the meninges, by the special fluid which separates these two layers, and by the fused bones of the cranium, its appearance gives no clue to its complicated working and functions.

The brain and spinal cord are among the first organs to develop when we are tiny embryos inside our mothers. Before the embryo is three weeks old a strip of tissue along its back has sunk in, the edges have rolled over, and a tube has been formed. This tube swells at the future head-end to become the brain, and the rest of the tube forms the spinal cord.

Both the brain and spinal cord are made of two types of nerve tissues: gray matter, made from cell bodies, and white matter, made from nerve fibers (see pages 10-11). The outer region of the brain is gray matter, the inner regions, white. In the spinal cord, white matter encloses the gray.

The brains of all vertebrates, that is animals with backbones, develop in the same way, and all vertebrate brains have the same major regions. Each region has its own role in the interpretation of information. How developed each region is depends on how complex the animal is and its place on the evolutionary ladder. Mammals, and human beings in particular, have the best developed cerebral hemispheres, as these regions are concerned with thinking behavior, or intelligence.

The vertebrate brain is divided into three main parts: the fore-, the mid-, and the hindbrain. In the

Below: The brains of animals vary widely in size. The smaller the brain in proportion to the skull, the less intelligent the animal is. The human brain is more developed than that of any other animal.

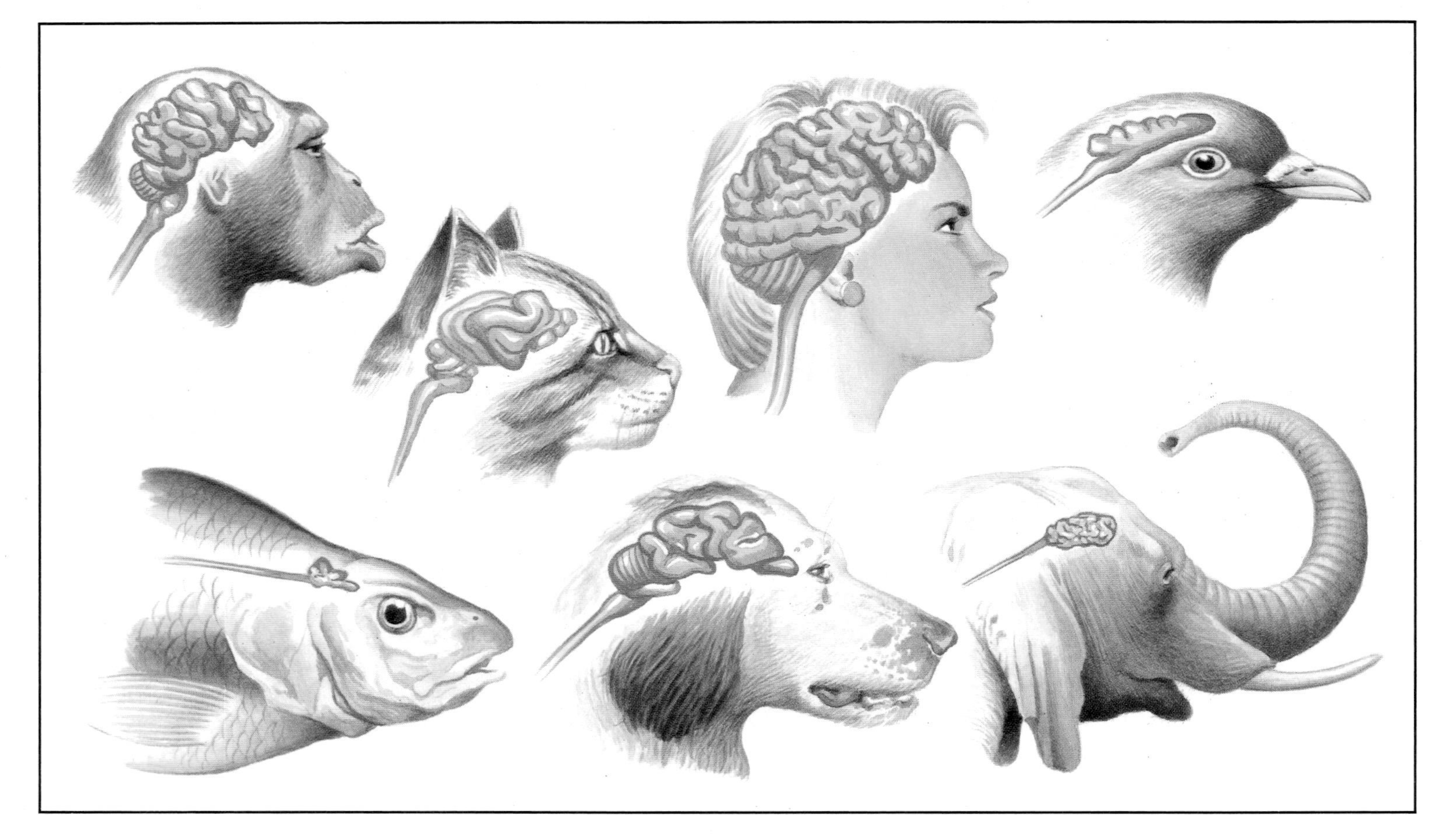

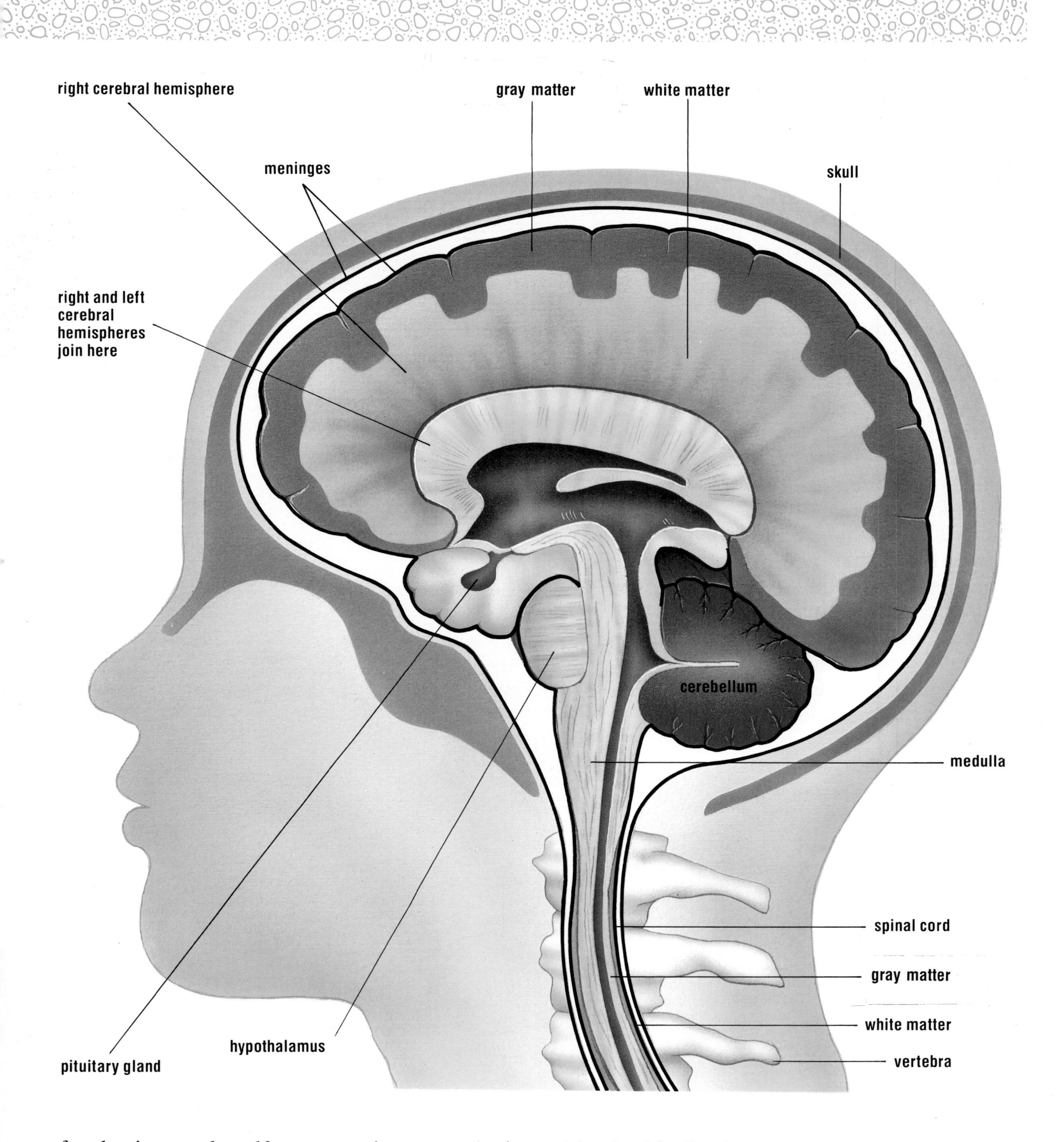

forebrain are the olfactory regions (concerned with the sense of smell), the cerebral hemispheres, and also the pituitary gland (see pages 32-33). In the midbrain are the optic lobes (concerned with seeing), and in the hindbrain are the cerebellum and medulla. The cerebellum coordinates all movement, and the medulla is concerned with heartbeat and respiration and is therefore crucial.

Above: The huge human brain fills much of the skull. The largest part of the brain is the cerebral hemispheres. These are concerned with intelligence and thinking.

THINKING AND FEELING

The human brain is the most advanced of all mammalian brains. Its most important and obvious development is the tremendous growth of the cerebral hemispheres, the intelligence center. These hemispheres have expanded so much that they cover most of the rest of the brain. In fact, the human brain is so large that the head of a baby is out of proportion with the rest of its body at birth in order to contain this tremendous development.

Different regions of the cerebral hemispheres carry out different jobs. Perhaps one of the most remarkable regions is the front part of the hemispheres, which is concerned with thinking. It is the ability to receive information, judge it against what we have either read or heard, and then come to a decision that makes the human being such an advanced animal.

LEFT AND RIGHT HALVES

The extraordinary development of our cerebral hemispheres conceals the fact that the human brain, like that of all other vertebrate animals, consists of two halves sitting side by side. Each half has the same regions and basic functions. The left half receives information from, and controls, the right half of the body, and vice versa. So, for example, if you move your left leg, it is the right half of your brain which has told you to do it.

However, the two halves of the brain also seem to work in different ways. The right side of the brain is concerned with artistic and creative feelings and activities; the left half seems to deal with mathematical and logical processes.

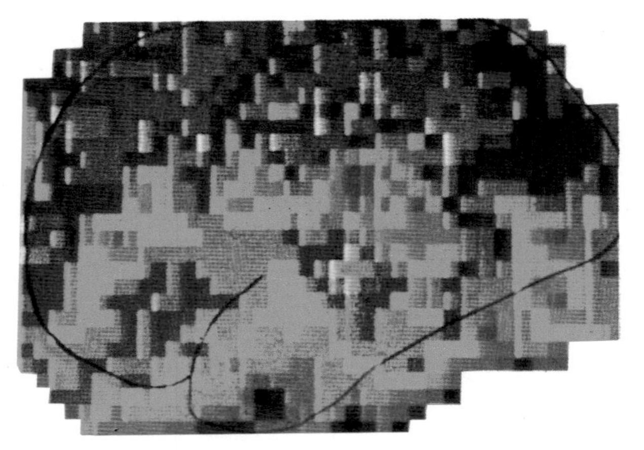

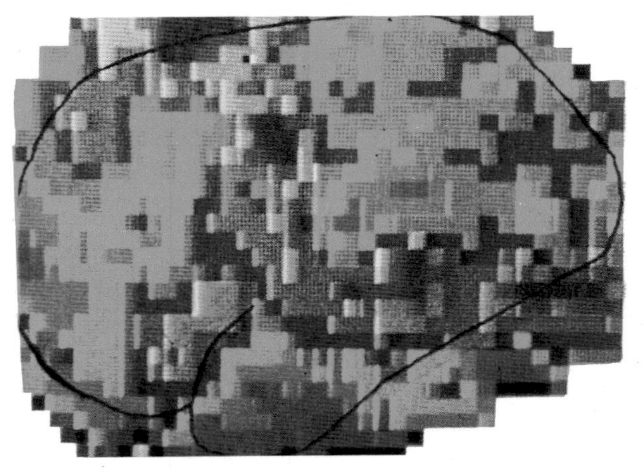

Left: Two scans of brain activity. The top picture shows the brain of someone resting, the lower of someone reading aloud. The yellow and red areas indicate where blood is flowing through the brain. The flow is very much faster when the brain is active.

Above and right: To play music we use mainly the creative right half of our brain, while to operate a computer, the more logical and scientific left half is important. However, to be really good at either activity means using both halves together.

Right: The two cerebral hemispheres of the human brain. Each of the color-coded areas controls a different activity (see key below).

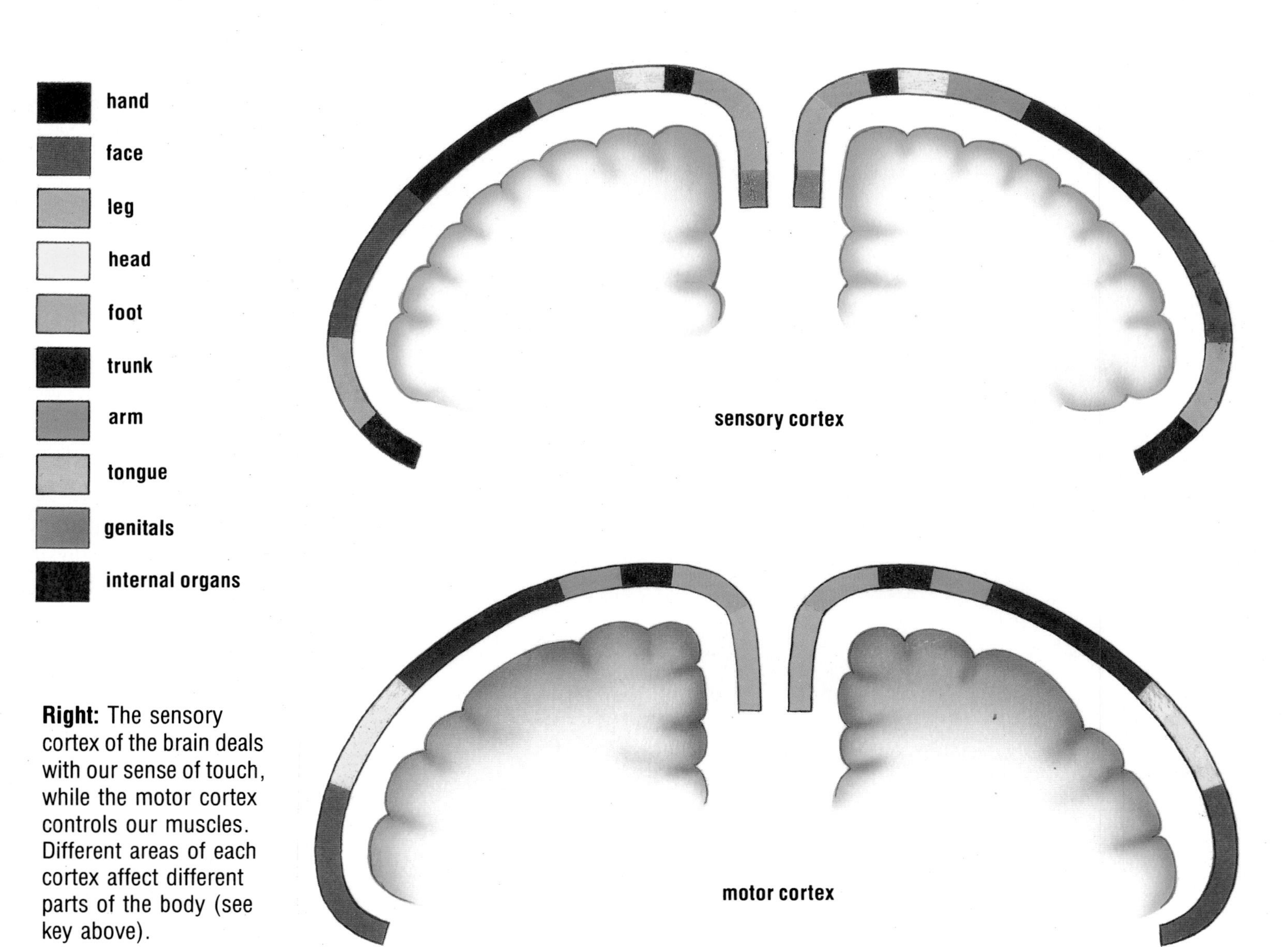

Right: The sensory cortex of the brain deals with our sense of touch, while the motor cortex controls our muscles. Different areas of each cortex affect different parts of the body (see key above).

HORMONES

The endocrine system, which produces all of the body's hormones, works alongside the nervous system as the other main control system of the body. Nerves operate using electrical impulses, some of which we can start and stop, while others are automatic. The endocrine system works automatically and we cannot control it, although our behavior may cause it to operate.

Imagine what would happen if you stepped out into a busy road — you would get a fright and with luck jump back out of harm's way. Meanwhile, the fright hormone adrenaline would have been automatically released into your bloodstream, causing a wide range of effects, including increased heartbeat and pupil size.

Below: A fairground ride may make us excited and slightly frightened at the same time. This will cause adrenaline to enter into the blood, making our heart beat faster and our liver release sugar, giving us energy.

Hormones are very complex chemicals produced by patches of cells called glands. The glands manufacture special chemicals and discharge them either directly into tissue or the blood, or else into a drainage duct. If the chemical leaves via a duct, the gland is called an exocrine gland. The sweat glands are exocrine glands. If the chemical is secreted directly into tissue or the blood, it is an endocrine or ductless gland. Hormones are the messengers produced by endocrine glands.

Because they are very powerful chemicals, only a little of each is produced at the time it is required. As they are carried all around the body, in the bloodstream, they pass through numerous organs. However, they are specialized to cause an effect in only very specific parts of the body.

A downgrowth from the brain called the pituitary gland controls all the other endocrine glands with its very special hormones. It receives information from all other parts of the body via part of the brain called the hypothalamus.

The pancreas is most unusual because it does the job of an exocrine *and* an endocrine gland. Positioned beneath the stomach, it produces chemicals which drain into the gut via a pancreatic duct, and it also produces a hormone, insulin. A shortage of insulin results in sugar diabetes.

The adrenal glands, above the kidneys, produce a range of hormones, including the very powerful adrenaline hormone that prepares the body to fight or run away and those that monitor blood minerals and mineral salts.

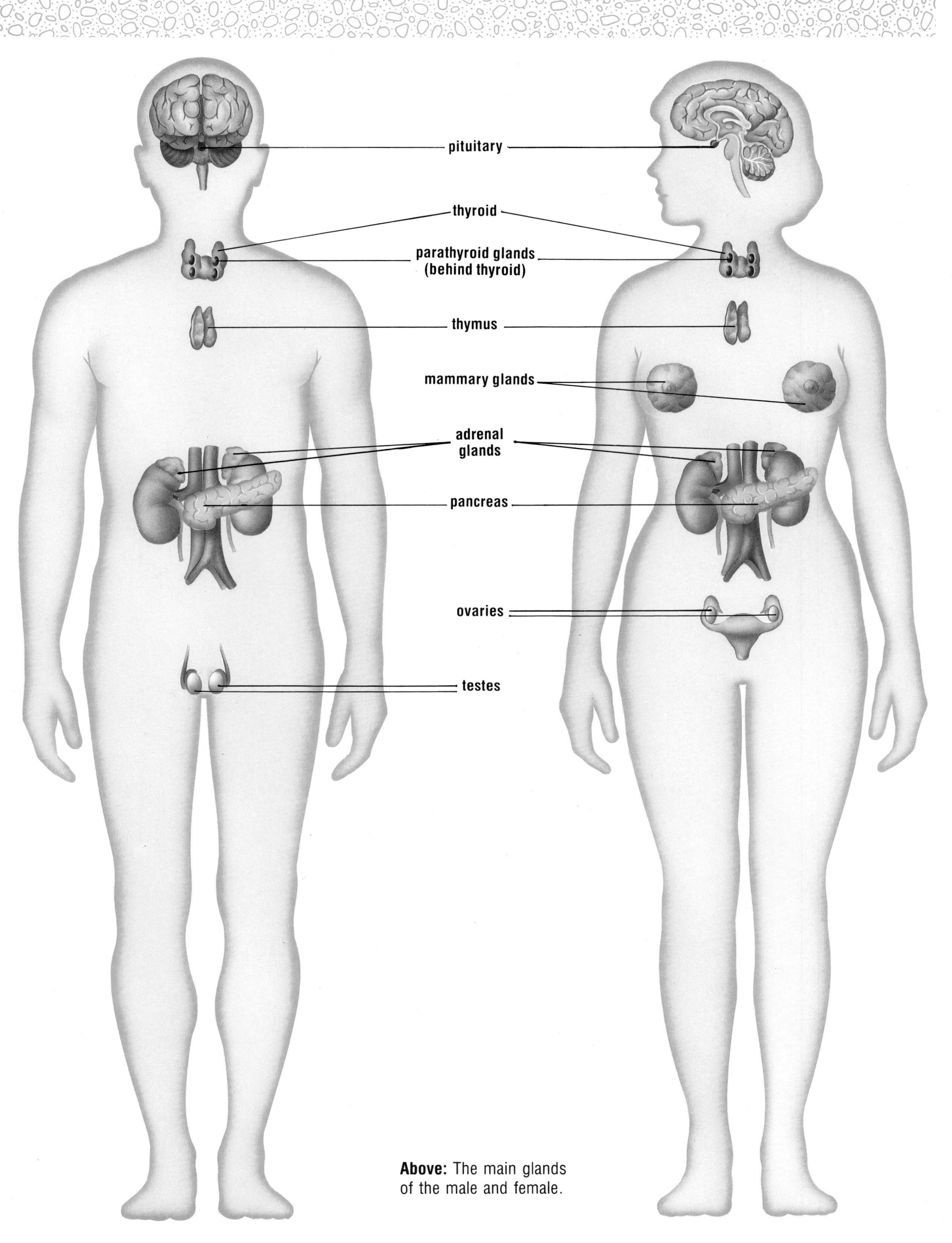

Above: The main glands of the male and female.

BEHAVIOR

Below: If we step on a sharp tack, we automatically remove our foot from the source of pain. This response is a reflex action. The nerve impulse caused by the pain passes via a nerve cell to the spinal cord. The brain is informed, but the message to move the foot is sent directly from the cord itself.

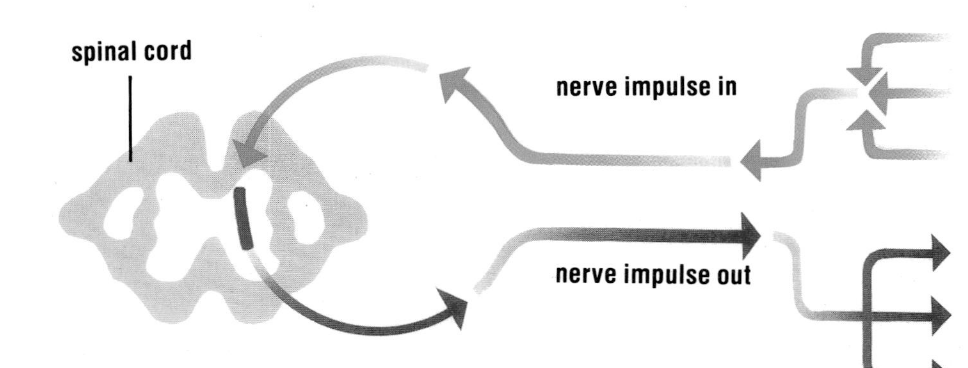

Our behavior patterns are the most complex in the animal kingdom. We have complicated lives that are ruled both by the laws of the land — our "outer" life — and by numerous changes in our bodies and minds which affect us very personally, what we call our "inner" life.

Much of our outer behavior we learn as children, either by watching adults or at school. There is no other animal that spends so much of its life learning from previous generations. Learned behavior is deliberately encouraged by the family and is the result of voluntary mental decisions and control. However, the way we live our inner lives is decided by a combination of involuntary body chemicals controlling us and voluntary action on our part.

To test that statement for yourself, think, for example, about a unicyclist. By choosing to display that skill to the public, he or she has made a voluntary choice. Once the decision is made, the cyclist will almost certainly become either anxious or excited, perhaps both. This causes the hormone adrenaline to flow — an automatic response by the body chemistry to a decision made by the mind.

All the time he or she pedals, the cyclist's body will be making thousands of tiny corrections to maintain good balance. These adjustments must be made very quickly, and so they are carried out by nerve impulses. The cyclist is not aware of this nervous activity and does not think about each action. In other words, these actions are automatic responses of the nervous system. They are what are called reflex actions, and they work alongside the automatic endocrine action.

The cyclist's reflex actions have been acquired by constant practice of a skill, but some of our reflex responses are within us at birth. Many of these inborn reflexes are protective, for example when we instantly pull our hands away from hot surfaces.

Right: If a newborn baby is allowed to fall backward, it will raise its arms and legs in the air. This action is called the Moro reflex. Babies also have other simple reflexes, such as blinking at lights.

Below: Sneezing is also a reflex action. It is caused by irritation of the lining of the nose.

COMMON REFLEXES		
Reflex	**Cause of reflex**	**Effect of reflex**
Production of saliva	Smell of food	Flow of saliva in the mouth ready to begin digestion
Sneezing	Irritation of lining of nose caused by dirt, germs, etc.	Contraction of muscles in the abdomen causing air to exit via the nose
Knee jerk	Tap on the knee tendon	Stretching of muscle and raising of lower leg
Coughing	Irritation of windpipe caused by dust, food, etc.	Contraction of muscles in the abdomen causing air to exit via the mouth

SENSES

We use five senses to keep in contact with the world about us — sight, hearing, taste, smell, and touch. Our bodies are constantly receiving stimuli which pass from the sense organs to the brain for interpretation. In some cases we are aware of what is happening, for example when we see a friend approaching and wave to him or her; in other cases, such as our automatic internal processes, we are unaware.

There are three types of sense organs. Firstly, the obvious ones which deal with messages from outside, for example the eyes and the ears. Secondly, there are specialized organs for detecting internal processes, such as the sugar concentration in the blood. The third type are those that keep us aware of the state of our muscles; these help us balance and also tell us how the muscles and tendons are performing, for example, if we pick up a heavy bag.

If a stimulus is applied over and over again, its effect lessens as time passes because the brain chooses to ignore the message. The pressure of eyeglasses on the head is one such example.

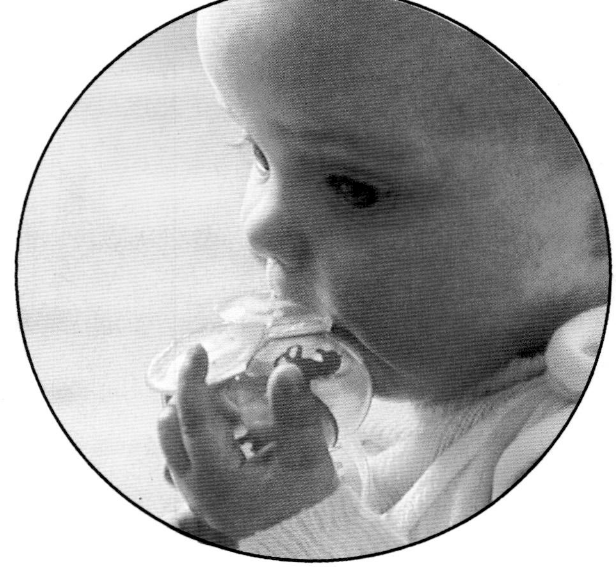

Left: Young children learn about the world around them using their five senses. The lips are an extremely sensitive part of the body and this baby is using them to feel the temperature and texture of the toy. The sense of touch is checked by using the sense of sight.

The same stimulus, such as the taste of a food, will set off the same impulses in us all, yet we will all respond differently, some liking the flavor, others not. It is this ability to respond in different ways that makes the study of human behavior so complicated.

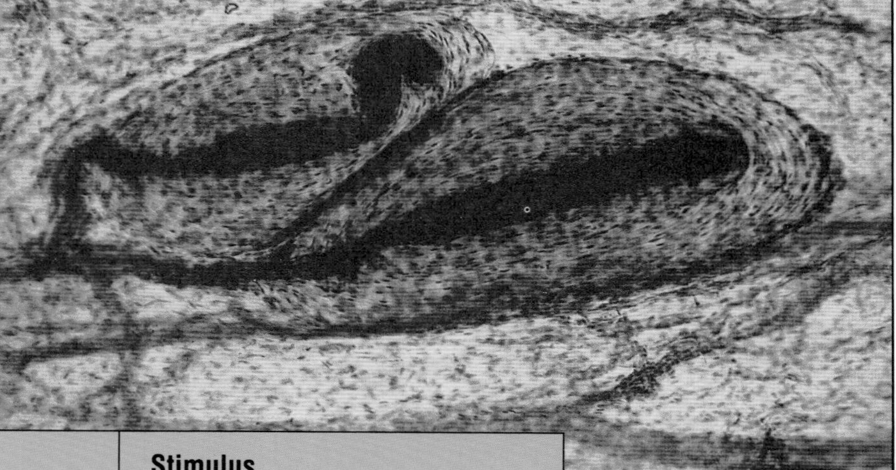

Above: In the dermis layer of the skin are five different types of nerve ending (see above right). This photograph shows a greatly enlarged nerve ending of the type sensitive to touch.

Sense organ	Function	Stimulus
Ear	Hearing	Sound
	Balance	Change in body position
Eye	Seeing	Light
Skin	Pressure detection	Touch
	Temperature detection	Heat and cold
Tongue	Tasting	Food and drink
Nose	Smelling	Chemicals in air

TOUCH

Our skin is very well equipped with sensory devices because it is the part of the body that is always in contact with the outside world. Although the sense of touch is found all over the body, certain parts, such as the lips, fingers, and toes, are more sensitive than others. Fingers, for example, are good at detecting even the slightest irregularity in a surface.

There are nerve endings just under the skin which detect faint pressure or irregularities, and others, more deeply buried, which respond when sharp objects dig in. If these deep nerve endings are stimulated, then the body feels pain. This is a safety mechanism telling us that whatever is causing the stimulus is dangerous.

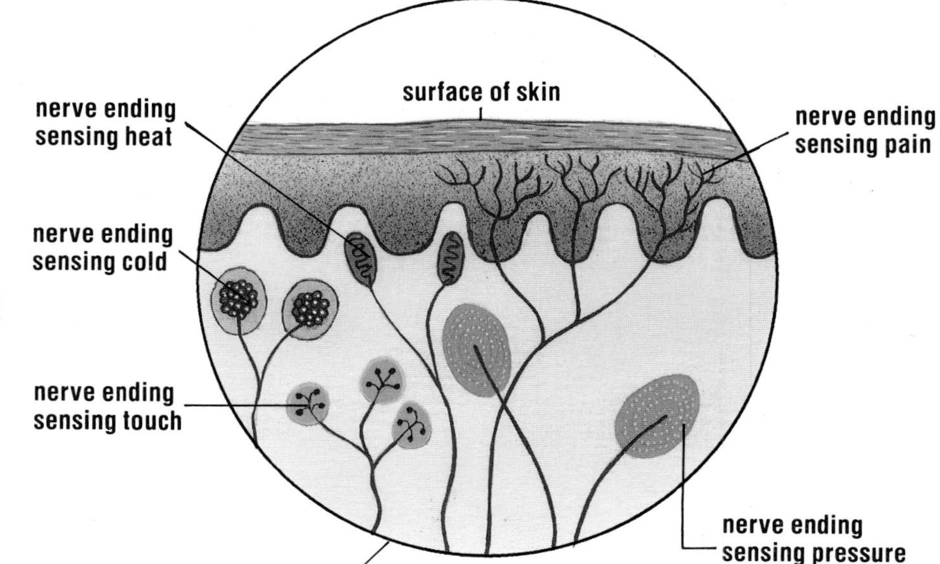

Above: The inner surface of our fingers is very sensitive. This is because it contains nerve endings sensing heat, cold, touch, pain and pressure.

Below: Hairs contain no nerve endings as they are made of dead cells. However, there are nerve endings around the follicles from which the hairs grow. This is why we can feel it if someone touches the hairs on our arms.

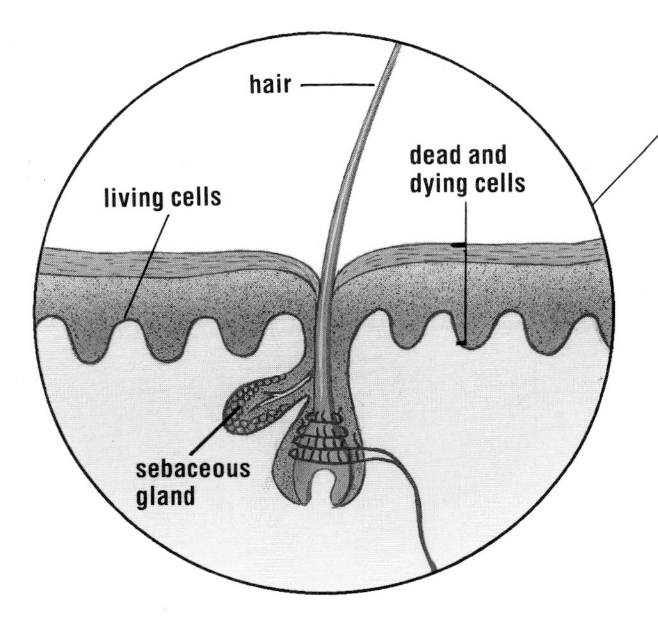

SEEING

At birth our eyes have almost reached their full adult size. They lie protected within bony sockets, positioned facing forward to give a wide field of view. Because they see objects at very slightly different angles, they add depth to what they see.

Eyes are held in place partly by the six muscles attached to each eyeball, and partly by the optic nerve which runs from the eye, through the back of the eye socket, and on to the brain. In addition the eyelids rest on the eyes, gently pushing on them. When you blink, each eyelid spreads a thin film of liquid, from the tear glands, over the eye. These tears contain a mild disinfectant to kill bacteria and fungi, and also a slippery solution to keep the eyeball clean and lubricated for movement.

The dark center, the pupil, is the window into the depths of the eye. The inside of the eye is not empty but filled with liquids. Behind the strong, transparent outer layer called the cornea, and in front of the lens, is a watery liquid, but filling the larger space behind the lens is a stiffer jelly. These materials help to keep the eye in its correct shape.

The colored circular band of the iris contains muscles and you can see these if you look closely into a mirror. They control the size of the pupil and so how much light enters the eye.

The lens is made up of a series of layers, like an onion, and it focuses to see distant or near objects by changing its shape. To receive light rays from a distant object, the lens stretches to become thinner; for near objects it becomes thicker.

Right: Many people need to wear eyeglasses to correct problems such as short-sightedness, when objects at a distance cannot be seen clearly.

Above and right: Rays of light from an object enter the eye through the pupil. These rays are bent by the cornea and then focused by the lens. An upside-down image is then formed on the retina. This image is passed to the brain by nerve impulses and we "see" the object. Of course the brain turns it the right way up!
Inset: The retina is made of special nerve cells called rods and cones.

The region that is sensitive to light rays is the retina at the back of the eyeball. The curved cornea and lens focus light onto the retina and the image formed is upside down and reversed! The sensory cells of the retina are the rods, which are concerned with black-and-white vision, and the cones, which enable us to see colors. There are three types of cones, each able to detect red, or blue, or green, which combine to give us all the colors we see.

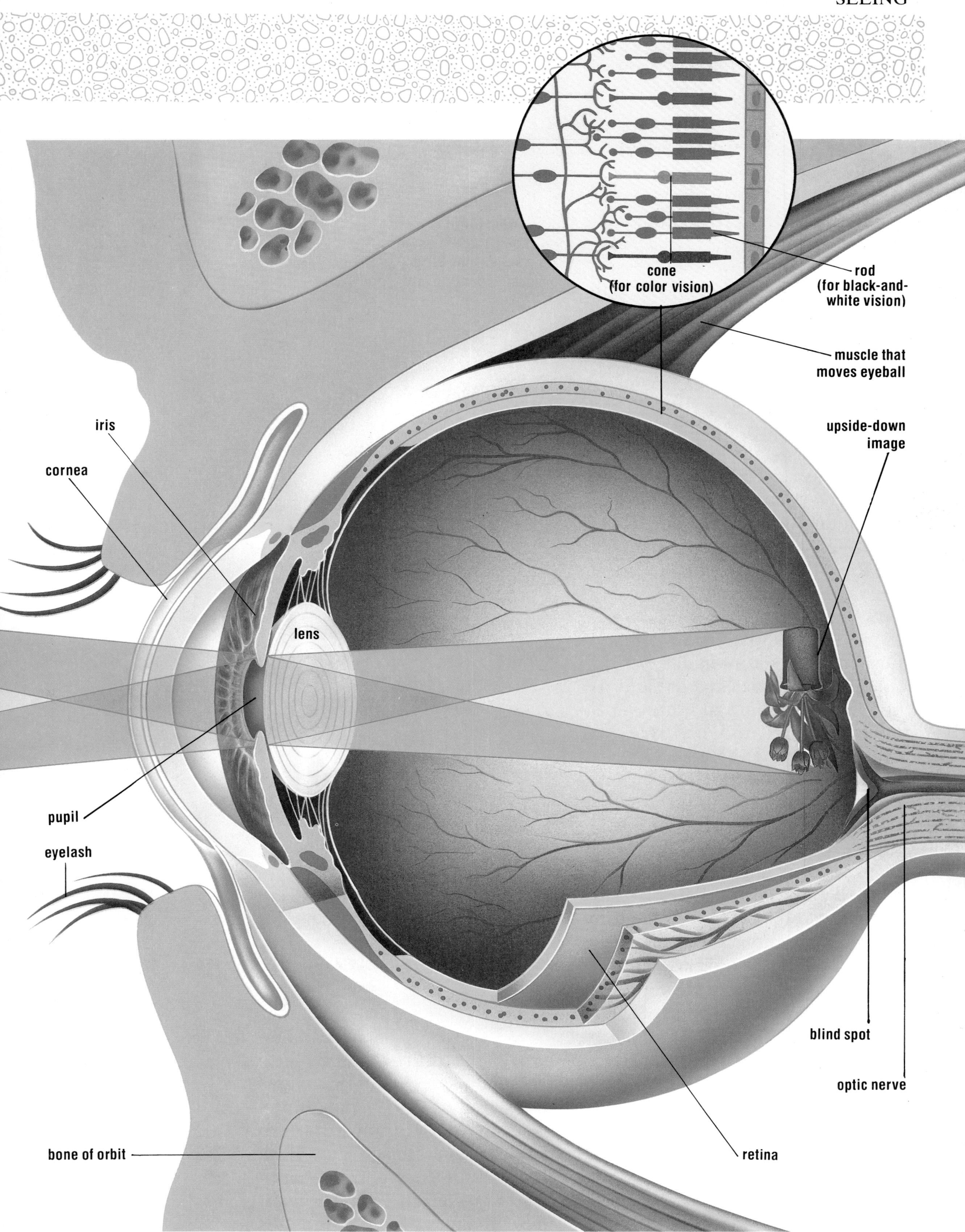
cone
(for color vision)
rod
(for black-and-
white vision)
muscle that
moves eyeball
iris
upside-down
image
cornea
lens
pupil
eyelash
blind spot
optic nerve
bone of orbit
retina

HEARING

The fact that we have visible external ears shows we are mammals, for no other group of animals has them. The human ear consists of three main regions: the external or outer ear that we see, the middle ear that connects with the back of the mouth, and the inner ear, made up of a series of tubes and coils.

The outer ear funnels sound waves into our heads through the ear hole. Having ears on opposite sides of our heads means we can immediately detect which direction a sound is coming from. Our ears are mounted in a fixed position, but many other mammals, such as the long-eared hare, can move their ears and even swivel them around.

Sound travels from the outside as far as the eardrum, which is pushed in by the pressure of the sound waves. The eardrum and the bony canal leading to it are kept moist with wax produced by special cells lining the ear canal. Occasionally too much wax is produced and it may collect over the surface of the eardrum. This can cause slight deafness because it stops full movement of the drum.

Movement of the delicate eardrum causes movement of the three tiny bones in the middle ear. These bones, called the ossicles, transmit the tiny vibrations of the eardrum across the air-filled space of the middle ear to a smaller thin skin known as the oval window. The Eustachian tube connects the air-filled middle ear with the back of the mouth and its function is to keep air pressure each side of the eardrum the same.

As they vibrate, the tiny bones or ossicles magnify every little movement of the eardrum so that the sound waves are just over 20 times stronger when they hit the oval window. Vibration of the oval window generates changes in the fluid within the coil of passages beyond called the cochlea, and these changes produce impulses in

Right: The delicate tubes, coils and bones of the ear are surrounded by solid bone. This protects them and keeps them in a stable position so that they can work properly.

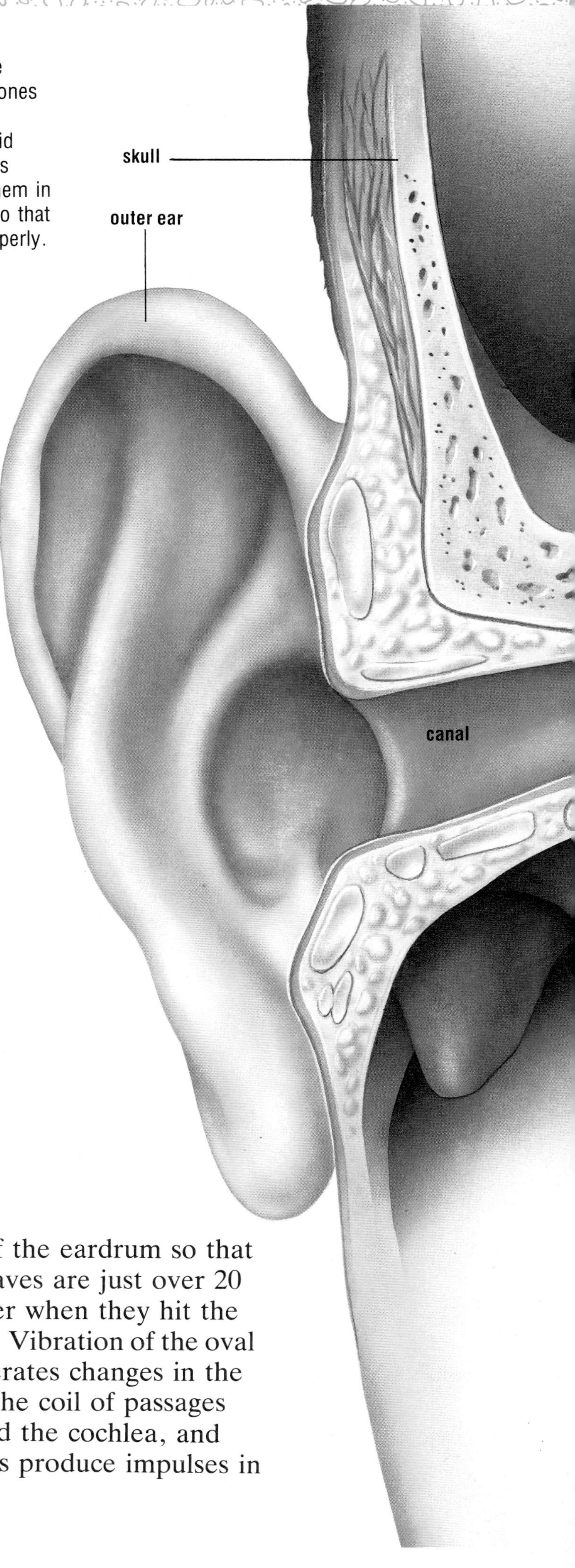

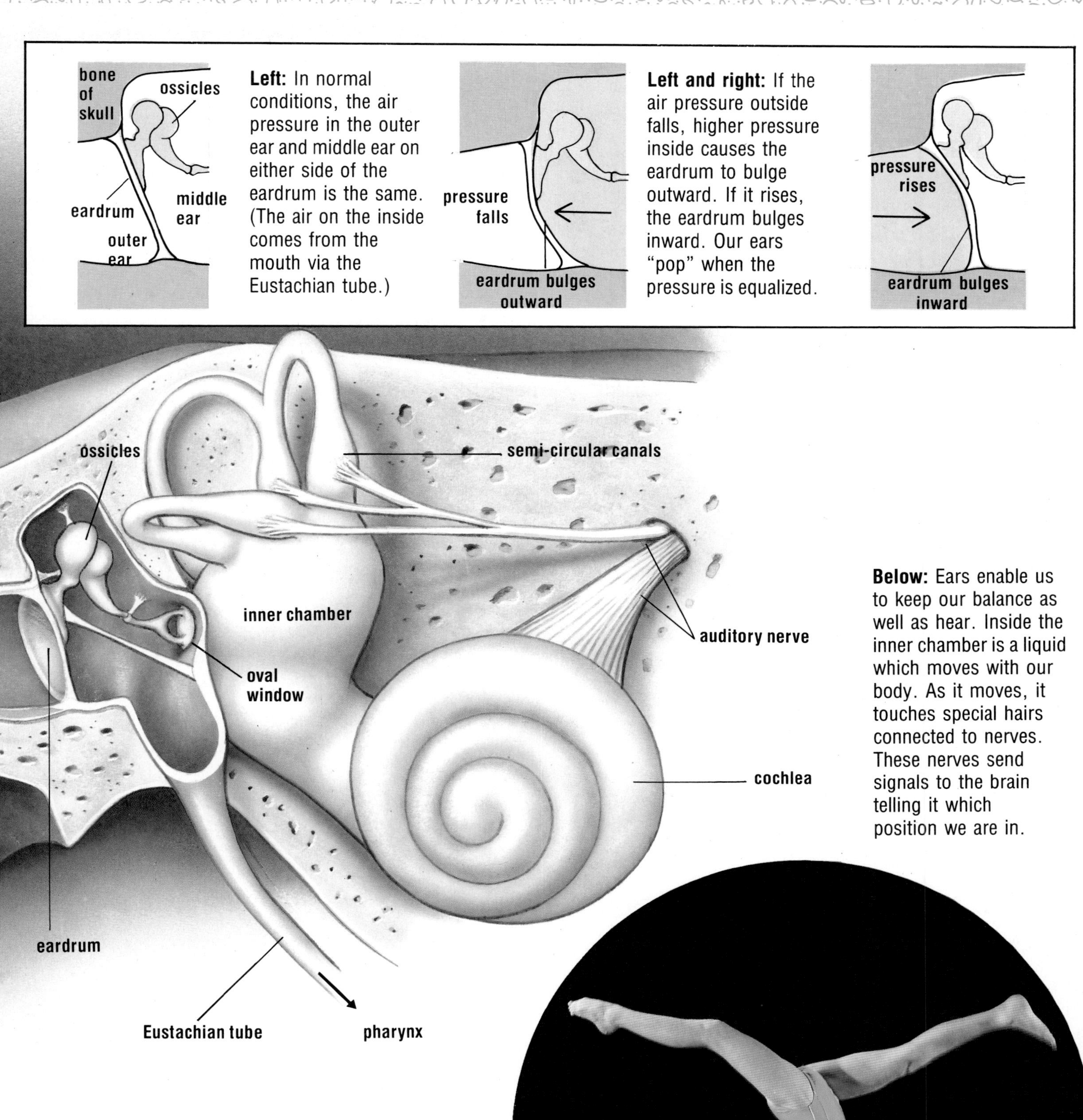

Left: In normal conditions, the air pressure in the outer ear and middle ear on either side of the eardrum is the same. (The air on the inside comes from the mouth via the Eustachian tube.)

Left and right: If the air pressure outside falls, higher pressure inside causes the eardrum to bulge outward. If it rises, the eardrum bulges inward. Our ears "pop" when the pressure is equalized.

Below: Ears enable us to keep our balance as well as hear. Inside the inner chamber is a liquid which moves with our body. As it moves, it touches special hairs connected to nerves. These nerves send signals to the brain telling it which position we are in.

the auditory nerve which the brain reads as sound.

Changes in general air pressure can also cause the eardrum to move — sometimes making ears "pop" and even resulting in pain. Really loud noise or sudden pressure change can burst an eardrum.

SMELL AND TASTE

There are two patches of scent-sensitive cells in the nose, sited in the upper part of the nasal cavity. These delicate cells are sunk in thin crevices and covered by a film of watery mucus. This liquid comes from special mucus cells and has two important jobs in the detection of smells. Firstly, it stops the sensitive cells from drying up, and, secondly, it provides moisture in which minute particles of the chemical which makes up the scent may collect.

When you sniff to find out more about a smell, you draw air, plus the particular chemical, up into the special crevices where the sensitive cells are to be found. Normal breathing in does not suck air into these uppermost creases of the nasal cavity.

The human sense of smell is much poorer than that of many animals (of dogs, for example), yet we are still able to detect a wide range of smells. Some people even rely on their noses for their employment, such as those whose job it is to tell the difference between various scents in perfume-making. However, in spite of the wide range of scents we can detect, our sense of smell tires very quickly. Often you find a smell that was unpleasant on entering a room quickly seems to fade.

Below: When we sniff something which has a strong smell, air containing the chemical causing the smell is drawn up into the nasal cavity. There, the chemical is dissolved in mucus and detected by scent-sensitive cells.

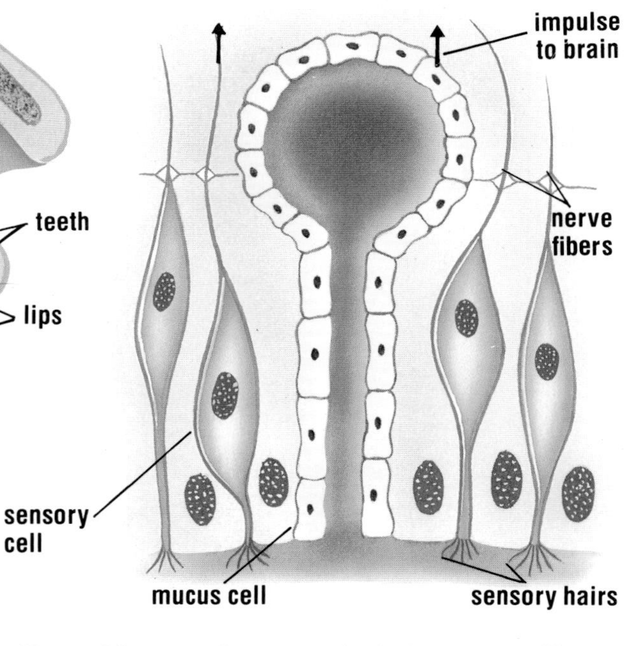

Above: The top of a crevice. The sensory hairs which detect smells are linked to the brain by nerves. The brain interprets the message it receives to identify the smell.

Right: The sense of smell is extremely important in the enjoyment of food. Usually we smell food before we taste it. This causes saliva to flow, which helps us to taste and digest what we eat.

Below: A false-color photograph of the tongue. The surface is very bumpy, so that there is a large area for the food to touch. Taste buds in the tongue detect the flavors.

TASTE

The sense of smell is very closely connected to that of taste. The sensitive cells in this case are collected into clumps, the taste buds, on the tongue, and different parts of the tongue detect different flavors. Some tastes are mixtures of the four basic types we can detect — bitter, sour, sweet, and salt — and therefore stimulate more than one area of taste buds.

If the nose is partly blocked, for example when you have a cold, your sense of taste is seriously affected. This means that to get the full flavor of some foods, the taste buds in the mouth have to be stimulated, while the scent drifts up into the nose.

Below: Different regions of the tongue detect different tastes.

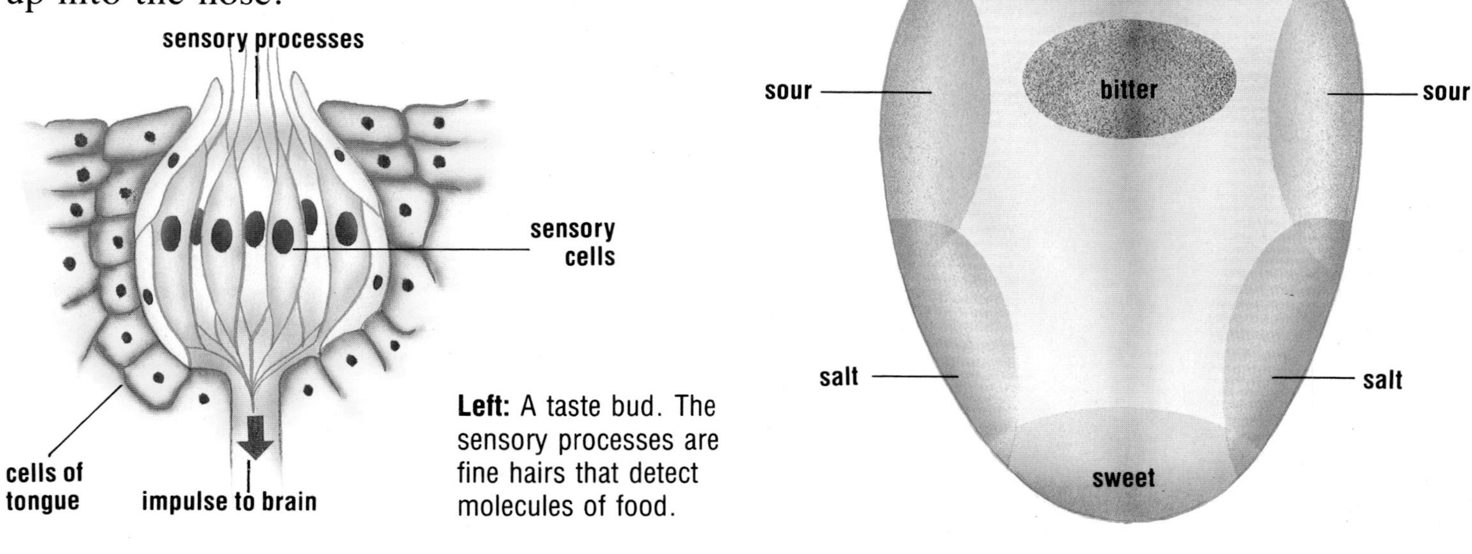

Left: A taste bud. The sensory processes are fine hairs that detect molecules of food.

TALKING

Two tubes run from the back of the mouth and down the length of the neck. One is called the esophagus and carries food to the stomach to begin the process of digestion (see pages 48-49), and the other is the trachea, or windpipe, which carries air to and from the lungs. At the top of the trachea is a tubular box, known as the larynx or voice box.

You can feel the front of your larynx if you very carefully and gently squeeze with your thumb and finger at the top of your neck, just where it joins the lower jaw. The larynx, or Adam's apple as some people call it, is supported by several pieces of cartilage and moves when you swallow.

Swallowing causes another flap of cartilage, called the epiglottis, to close over the top of the windpipe. This prevents saliva and food from passing into it. Occasionally that does still happen, especially if you try to talk and swallow at the same time, but your body reacts by coughing, in other words forcing air up the windpipe to blow the offending material out. This is what happens when something "goes down the wrong way."

Below: Choirboys have beautiful, high voices. As they grow older their larynxes enlarge and their voices "break" and become deeper. Some singers still have good voices after this, but others do not.

MAKING SOUNDS

The larynx is lined with a smooth, moist layer of skin. Two folds of tissue extend from the lining across the larynx toward each other, leaving a narrow gap between them. These folds are the vocal cords and air must pass through the gap between them on its way to and from the lungs. As the air passes up between the vocal cords it makes them vibrate, producing a sound. This sound travels up to the pharynx, a moist muscular chamber

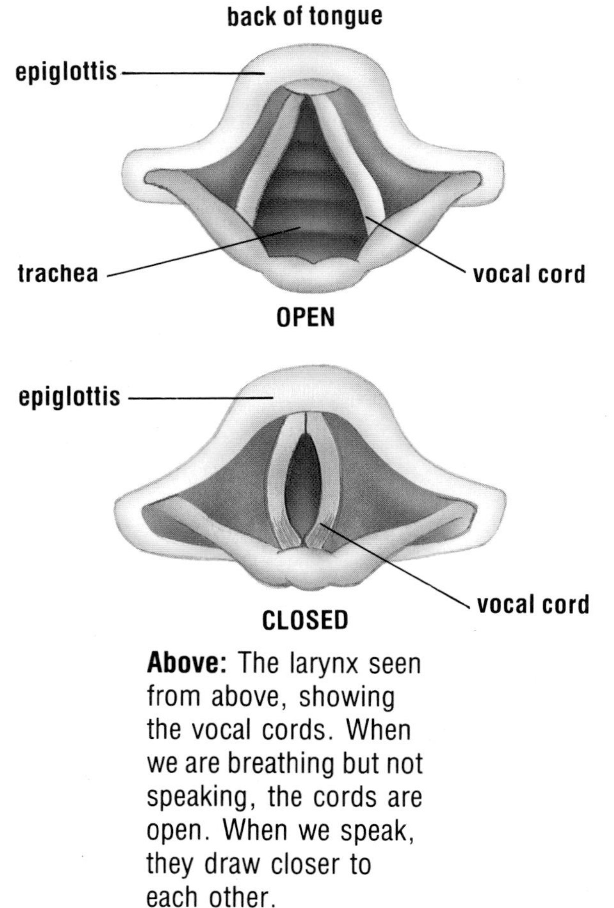

Above: The larynx seen from above, showing the vocal cords. When we are breathing but not speaking, the cords are open. When we speak, they draw closer to each other.

at the back of the mouth, which acts as a sound box.

The type of sound you make depends on how tightly the vocal cords are held by the muscles of the larynx. Taut vocal cords produce high-pitched sounds, slack cords, deeper notes. Place your fingers very gently on your larynx and say your name and address, varying the high and low notes. You will feel various different vibrations as you speak.

Men have deeper voices than boys because as boys grow up the larynx enlarges slightly so that the voice "breaks." Girls' larynxes do not grow so much and therefore their voices change little.

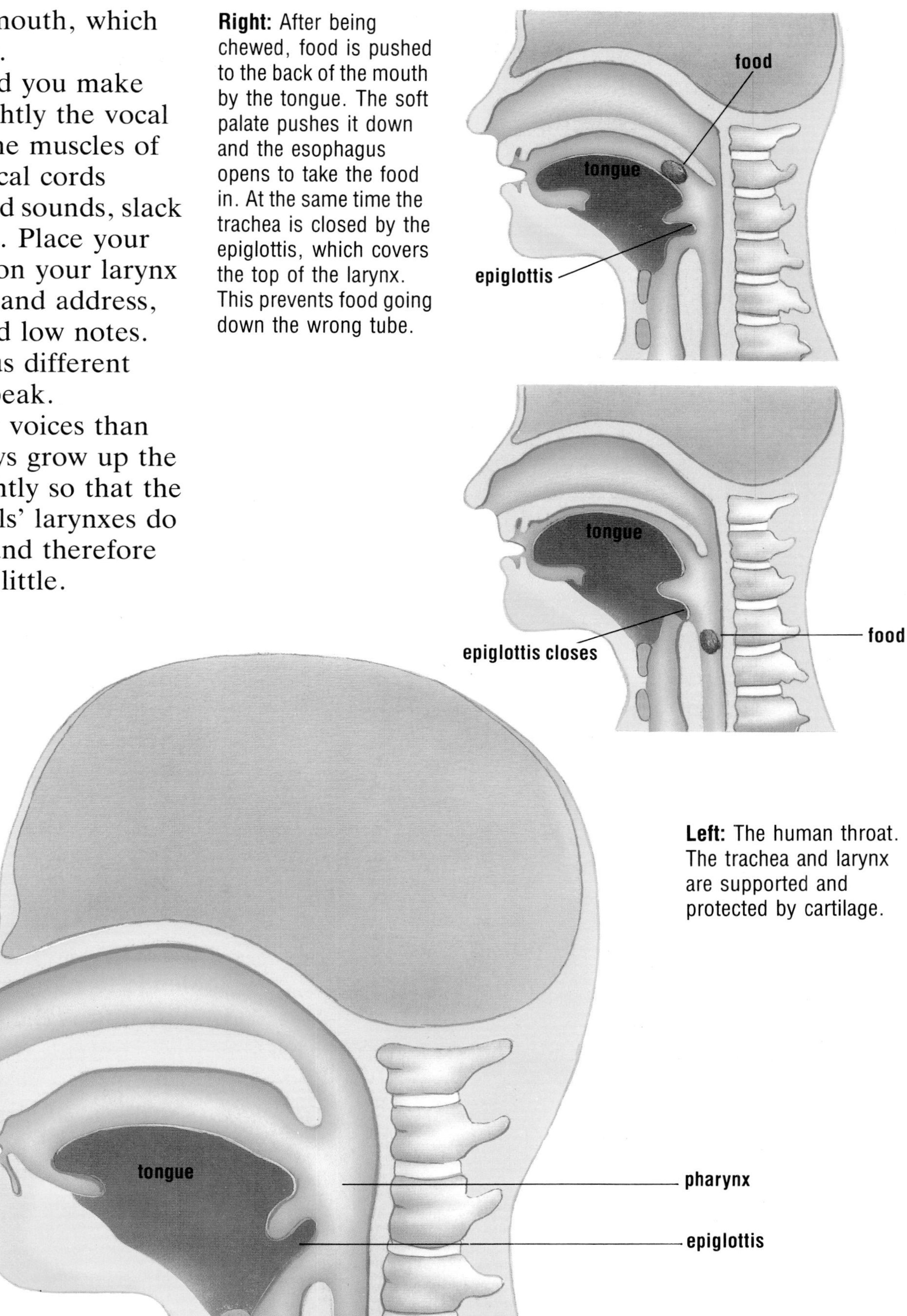

Right: After being chewed, food is pushed to the back of the mouth by the tongue. The soft palate pushes it down and the esophagus opens to take the food in. At the same time the trachea is closed by the epiglottis, which covers the top of the larynx. This prevents food going down the wrong tube.

Left: The human throat. The trachea and larynx are supported and protected by cartilage.

TEETH

All human teeth grow from sockets in the jaw and thus have a root region, embedded in bone, and an upper section, or crown, the part we can see.

Humans have two sets of teeth — the juvenile set, or milk teeth, of which there are 20, and the permanent adult set of 32. The adult set grow up through the milk teeth, which by the age of about six years have started to fall out. Occasionally adults still have a few milk teeth, but usually they have all gone by the age of 14. Although we grow a second set of teeth, it is important to look after the milk teeth because decay can damage the growth of the adult set.

We use our front teeth, the incisors, to cut off pieces of food which the larger and heavier premolars and molars can then chew up. Chewing crushes the material and makes it easier to swallow. Our canines, the teeth behind the incisors, now do the same job as the premolars. However, many of our close relatives in the animal kingdom still use them for attack.

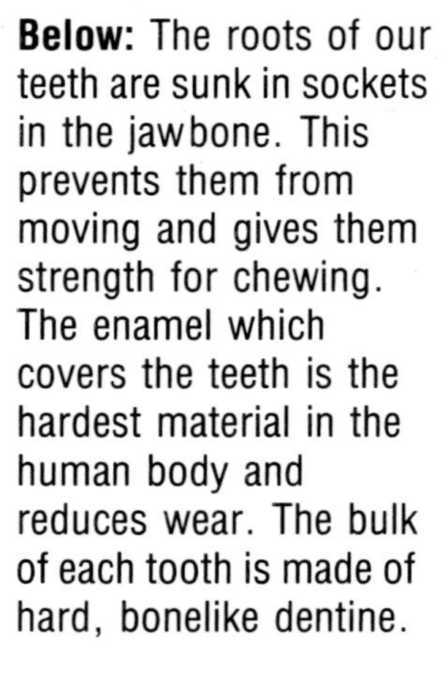

Below: The roots of our teeth are sunk in sockets in the jawbone. This prevents them from moving and gives them strength for chewing. The enamel which covers the teeth is the hardest material in the human body and reduces wear. The bulk of each tooth is made of hard, bonelike dentine.

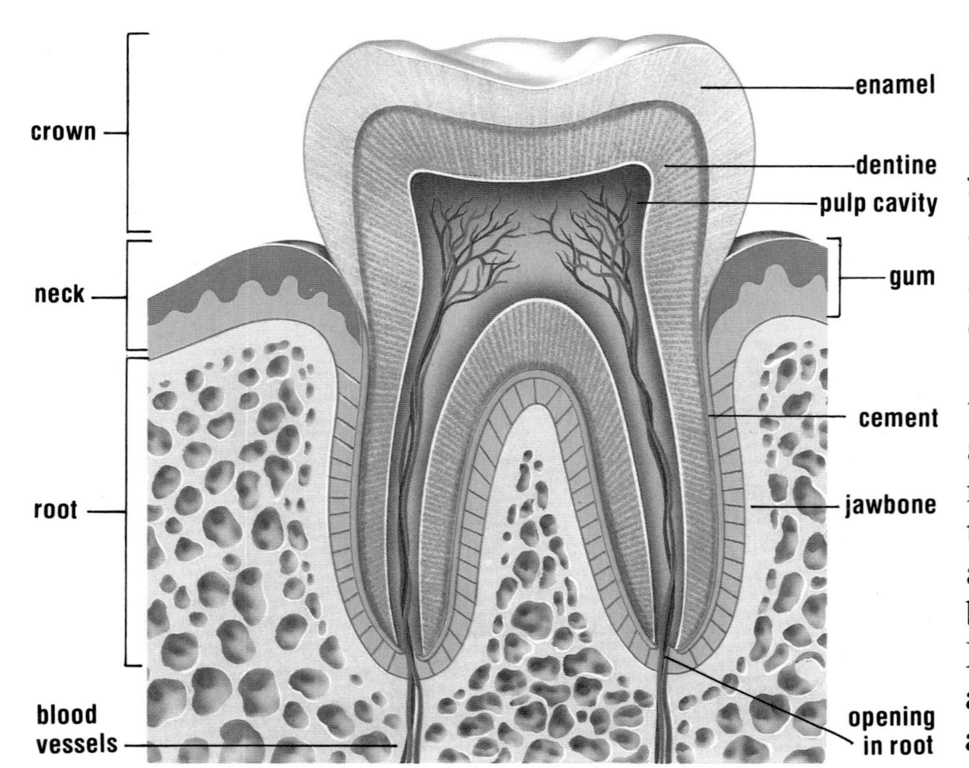

Chewing any food leaves small particles stuck between and on the teeth and these provide a food source for the bacteria that live in our mouths. This is particularly true if the food has sugar in it. The bacteria digest the sugar, and in so doing produce acids which attack the surface of the teeth and the skin of the gums. These acids can actually eat away the enamel of the tooth, especially if there are any minute cracks or scratches on it.

If not treated, the acid attack will continue through the enamel and dentine, until it reaches the pulp cavity containing the nerve endings. This produces the very unpleasant pain of toothache. Occasionally the bacteria and the chemicals they make infect the pulp cavity so much that they cause an abscess, which results in terrible pressure and pain.

If the gum is attacked by acid, it usually swells, becomes inflamed, or red, and may bleed easily. This leads to more damage at the junction of the tooth and its socket. The cement may be loosened, which can cause still more tooth decay or simply a loose tooth that eventually drops out.

Tooth decay can be reduced in two ways: by cutting down the amount of sugar you eat and, even more importantly, by brushing teeth very regularly, especially after meals. This prevents the build-up of a layer of bacteria. Regular check-ups at the dentist are also very helpful in stopping any damage from getting worse.

Below: In this child's jaw, the milk teeth are fully grown, while the adult teeth (shown in blue) are developing under the gums.

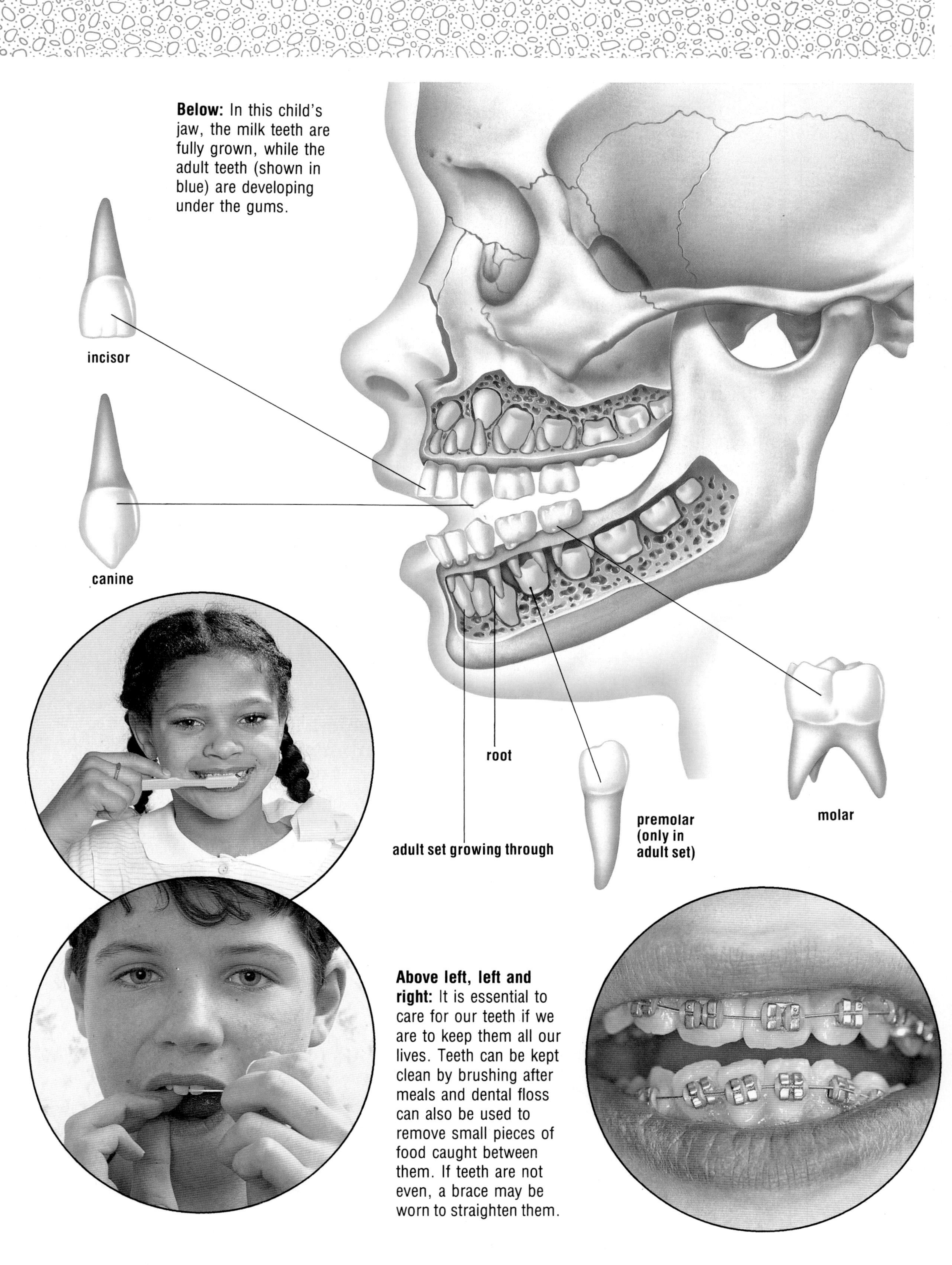

Above left, left and right: It is essential to care for our teeth if we are to keep them all our lives. Teeth can be kept clean by brushing after meals and dental floss can also be used to remove small pieces of food caught between them. If teeth are not even, a brace may be worn to straighten them.

DIGESTION

We need food and water to fuel all the complex jobs our bodies do to keep us alive. But to use that fuel, or energy, we need to break what we eat and drink down into materials our bodies can handle. Digestion is the breaking down of food into smaller pieces of simpler chemicals so that they are dissolved in the liquids of the intestine and then absorbed. Different foods need different treatments and the structure of the alimentary canal, or gut, plus various "helper" glands and organs allow these different events to take place in their own time.

The first stage is the attack on large pieces of food by the teeth. Biting off bits and chewing them makes the food softer and easier to swallow because it is mixed with saliva from the salivary glands. Swallowing food starts it on a long journey as it passes down the esophagus to the stomach. The process by which it is moved along the gut is called peristalsis and consists of a series of muscular squeezing actions.

CHEMICAL ATTACK

After the physical attack on food in the mouth, the rest of the process is chemical and the agents are various sorts of enzymes, that is complex proteins which are produced by the body.

The glands of the stomach produce dilute hydrochloric acid to sterilize food and also enzymes to start protein digestion — protein digesters work best in acid conditions — plus lots of mucous slime to protect the stomach wall itself from the acid and enzymes.

In the small intestine there is yet more mucus, plus a range of glands producing enzymes that attack carbohydrates (like starch and sugars), fats, and the remaining proteins. It is the job of the villi (see page 50) on the walls of the small intestine to absorb digested food, and any material left after this has no further value as food.

Therefore the large intestine has two main jobs — to absorb water in the colon and to store waste matter in the rectum until it is eliminated through the anus.

Right: Our food travels on a long and complicated path through our bodies. From the mouth, it first passes down the esophagus to the stomach. After the food is digested there, it travels into the small intestine. Enzymes from the pancreas, and bile produced by the liver and stored in the gall bladder, arrive here via ducts to help in the digestive process. Any remaining material moves on to the large intestine where water is removed from it. Finally it is stored in the rectum and passed out of the body through the anus.

Below: The digestive process starts when we bite off and chew pieces of food. Saliva in the mouth moistens the food so that it can slip easily down the esophagus.

Below: The gut wall has two layers of muscle. When the circular muscles contract, they squeeze food along. When the longitudinal muscles contract, they stretch the circular muscles back to their original shape. This squeezing action is called peristalsis.

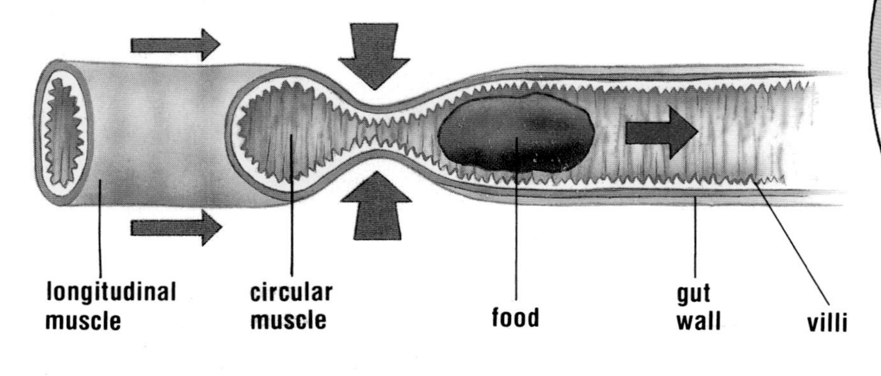

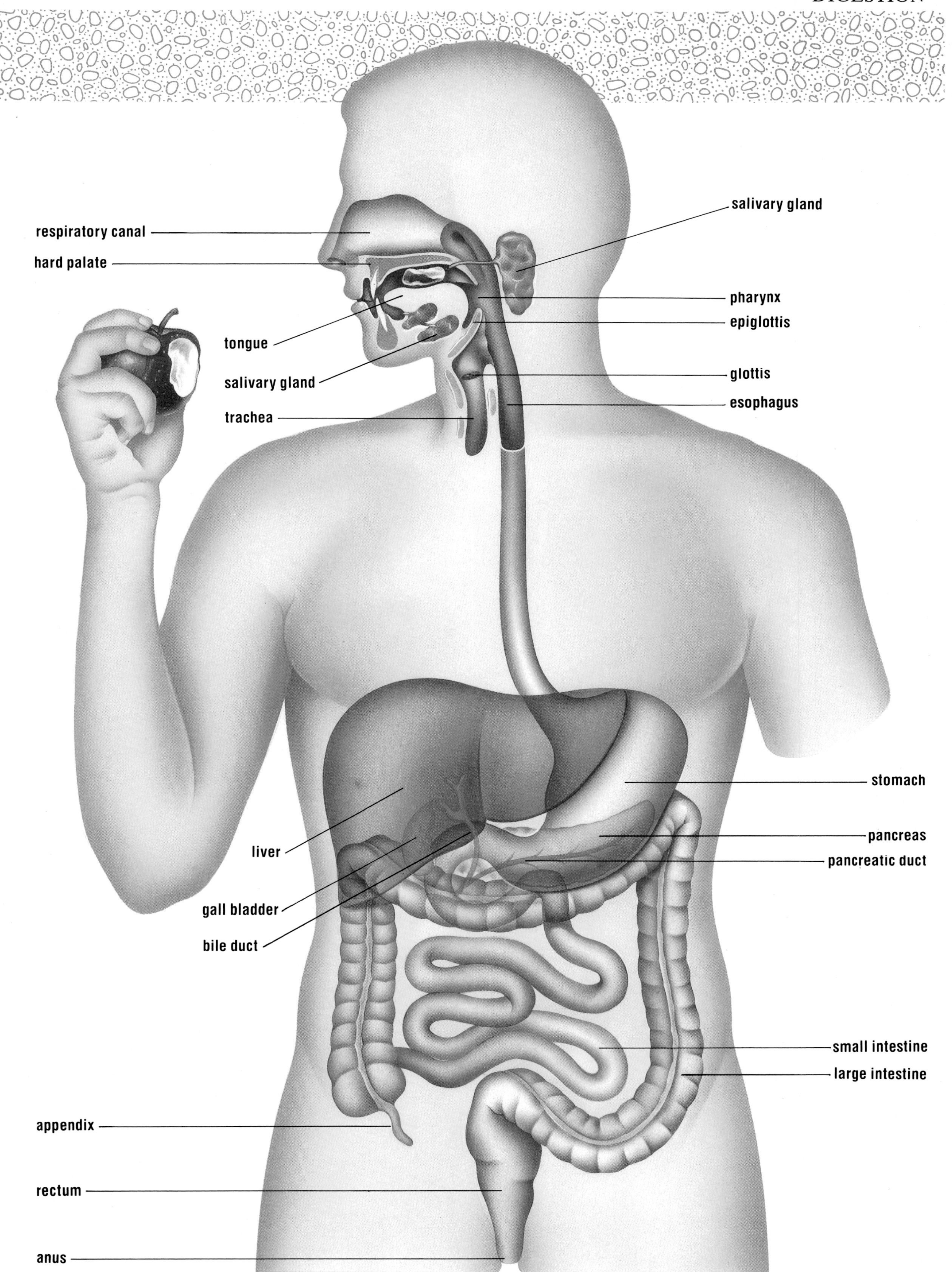
respiratory canal
hard palate
tongue
salivary gland
trachea
salivary gland
pharynx
epiglottis
glottis
esophagus
stomach
pancreas
pancreatic duct
liver
gall bladder
bile duct
small intestine
large intestine
appendix
rectum
anus

ABSORBING FOOD

By the time food gets into the small intestine, the large molecules making up the different materials have all started to break down. The resulting molecules are now in a form that can pass through the delicate lining of the small intestine and into the blood.

To make sure none of the small molecules are wasted, the surface area of the intestine lining has thousands of tiny finger-like outgrowths called villi. Together they add up to an enormous surface area through which we absorb sugars, amino acids (broken-down proteins), and simple fat products. A large number of minute blood vessels called capillaries run under the skin of each villus, collecting the sugars and amino acids and transporting them away. Fat products pass into the lacteal vessels running through each villus for transport by the lymphatic system (see pages 56-57).

The capillaries of the small intestine join to form a large vein, the hepatic portal vein, which carries food materials to the liver. The liver is responsible for checking the food content of the blood. If there is too much sugar, it is stored either in the liver itself, as a special compound, or taken to store as fat elsewhere — under the skin or around the kidneys, for example. But, by traveling in the lacteal vessels of the lymphatic system, fat bypasses the liver and joins the bloodstream at major junctions in the lymph nodes at the groin, neck, and armpits.

The hepatic portal vein also collects blood from veins draining the large intestine. The capillaries here collect liquids from the gut,

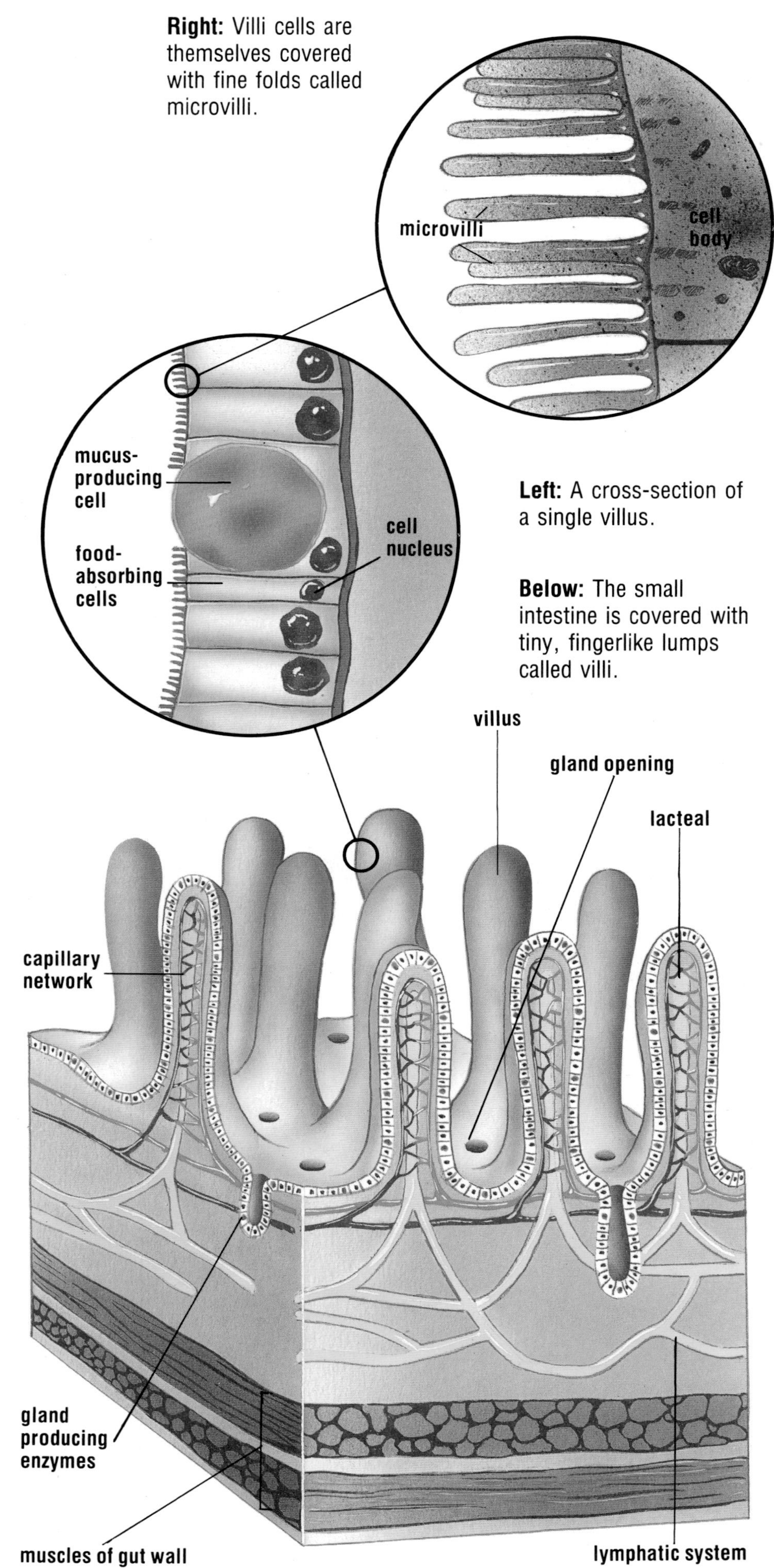

Right: Villi cells are themselves covered with fine folds called microvilli.

Left: A cross-section of a single villus.

Below: The small intestine is covered with tiny, fingerlike lumps called villi.

Right: Food is absorbed into blood vessels in both the small and large intestines. These vessels join up to form the hepatic portal vein, which carries the digested material to the liver, the largest organ in the body. The liver releases the digested food to the rest of the body as it is required.

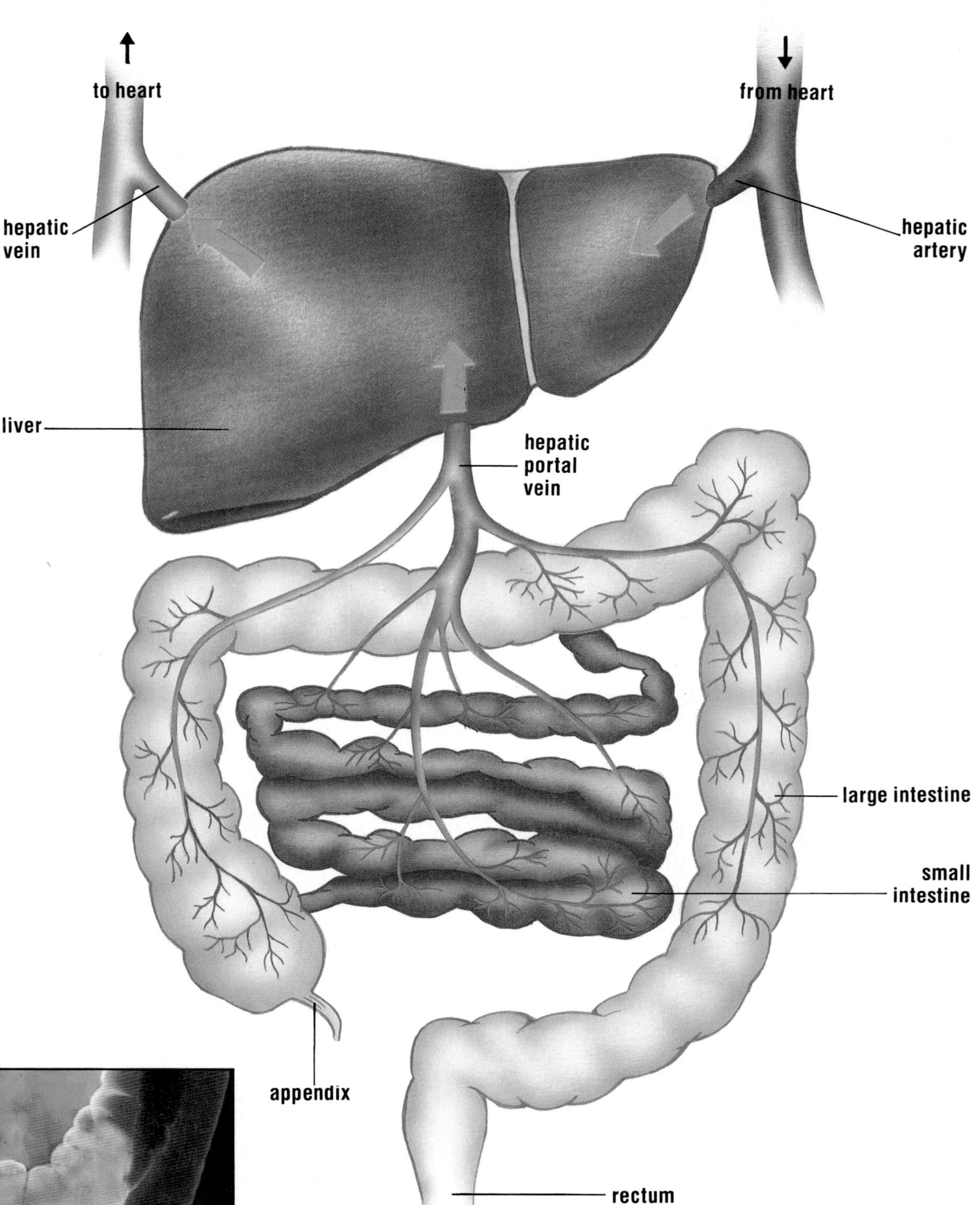

Below: A false-color X-ray of the large intestine. The large intestine is much wider than the small intestine, but also much shorter. It is 6 ft long, while the small intestine measures 21 ft.

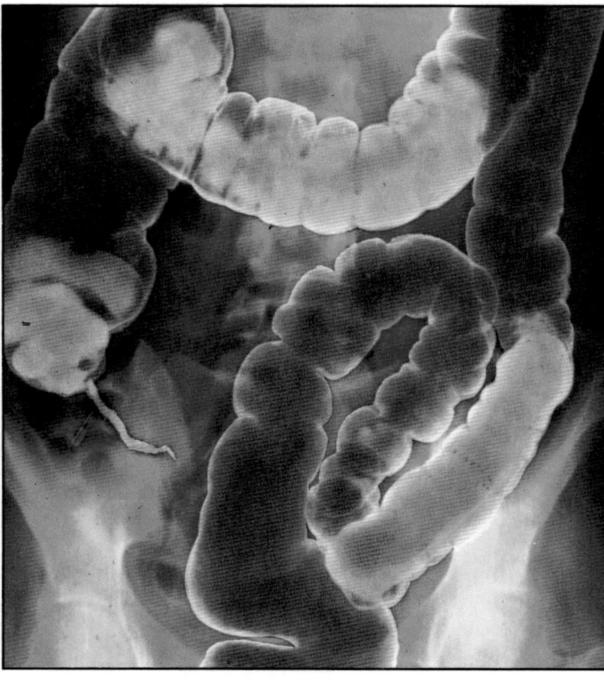

but its lining is simply folded and has no villi. As land animals, we need to conserve our water supply, and many of the secretions used earlier in the digestive process, for example mucus and enzymes, are mainly composed of water, so the colon reclaims it from what is by now waste food.

EXCRETION

Many of the processes going on in our bodies produce waste materials. This may be the result of cells and tissues wearing out, complicated chemicals like hormones and enzymes being broken down, or simply too much protein in our diet. Whatever the origin of the waste, if it contains the element nitrogen, it will be dealt with first by the liver and then by the kidneys. Chemicals derived from nitrogen, especially ammonia, are deadly poisons to the human body.

The liver is able to turn a wide range of chemicals containing nitrogen into simpler materials. Some of these simpler forms are still potentially dangerous and so they are combined with another waste material, carbon dioxide, to form a chemical called urea. Urea is then carried in the bloodstream from the liver to the kidneys to be filtered out.

Blood enters the kidneys under high pressure within the renal arteries. Inside each kidney, the renal artery divides again and again into smaller and smaller blood vessels, which finally form little knots of capillaries in the outer region of the kidney called the cortex. The blood pressure here is so high that it forces much of the water and dissolved materials out of the capillaries and into fine tubes called nephrons.

It is while the liquid trickles along the nephrons that useful materials are removed back into the blood. Chemicals such as glucose and water are reclaimed, leaving a little of the water plus the waste urea. This liquid is urine. Urine passes into the central cavity of the kidneys, from where it is

Far right: There are two kidneys in the human body, which lie one on either side, just behind the abdomen. Urine from each kidney travels down the ureter to the bladder, where it is stored until it can be passed out of the body.

Below: Babies cannot control the excretion of waste material from their bodies. This is why they must wear diapers. As they grow older, they are toilet-trained to learn this control.

Below: A kidney machine can be used to clean the blood of a person whose kidneys do not work properly.

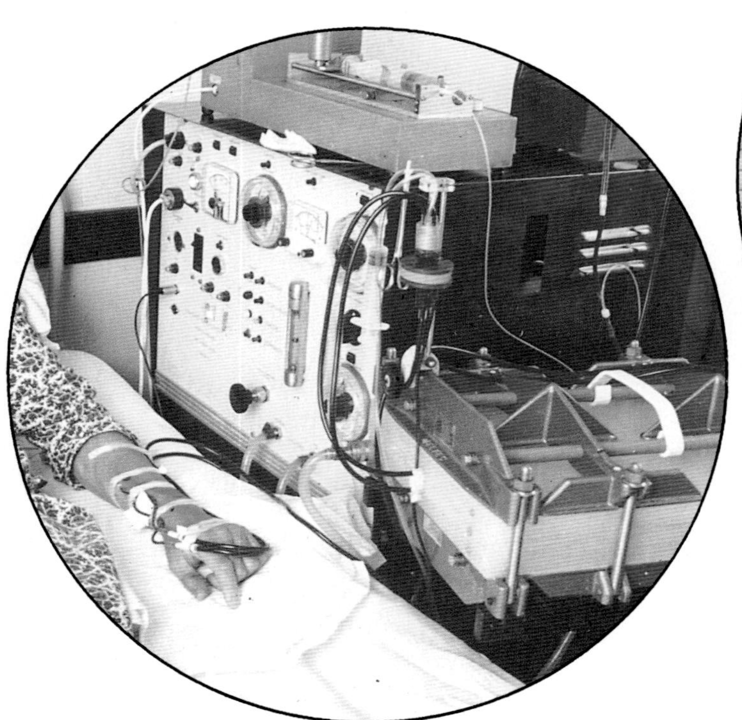

Right: Inside each kidney between branches of the renal artery are over a million tiny tubes called nephrons. In them, useful materials which have been forced out of the blood are reabsorbed.

carried down into the bladder for temporary storage. From time to time we voluntarily release urine, thereby relieving the pressure which builds up in the bladder.

The amount of water we lose as urine can vary. If it is hot, or we have been very active, we sweat to cool down. Because it is dangerous for the body to lose too much fluid, water is reclaimed from the nephrons back into the blood to replace what was lost through sweating. The amount of urine produced is therefore smaller, and it is more concentrated. Obviously, if we drink lots of fluids, we release lots of dilute urine.

The solid waste material stored in the rectum and passed out of the body through the anus is not strictly speaking part of the excretion process, being merely unused and indigestible food material.

to heart
from heart
liver
renal artery
right kidney
renal vein
left kidney
ureter
bladder

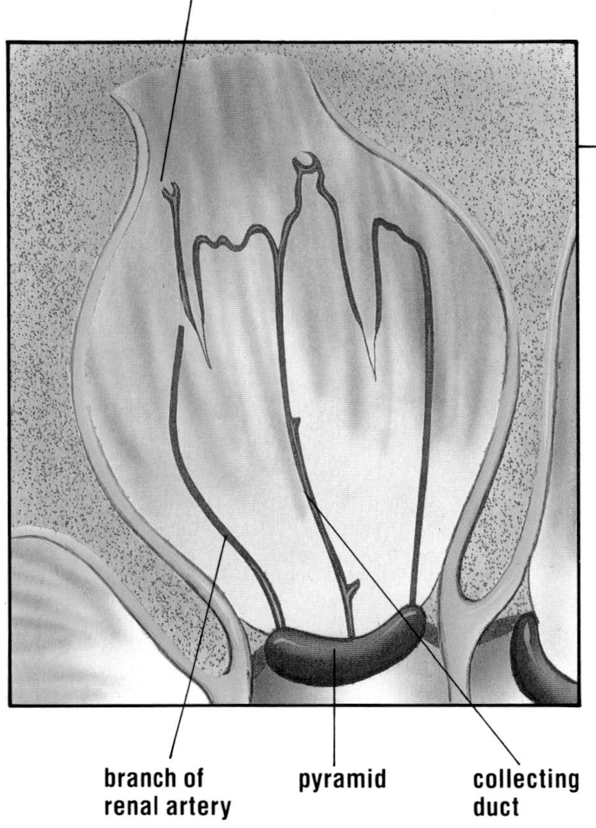

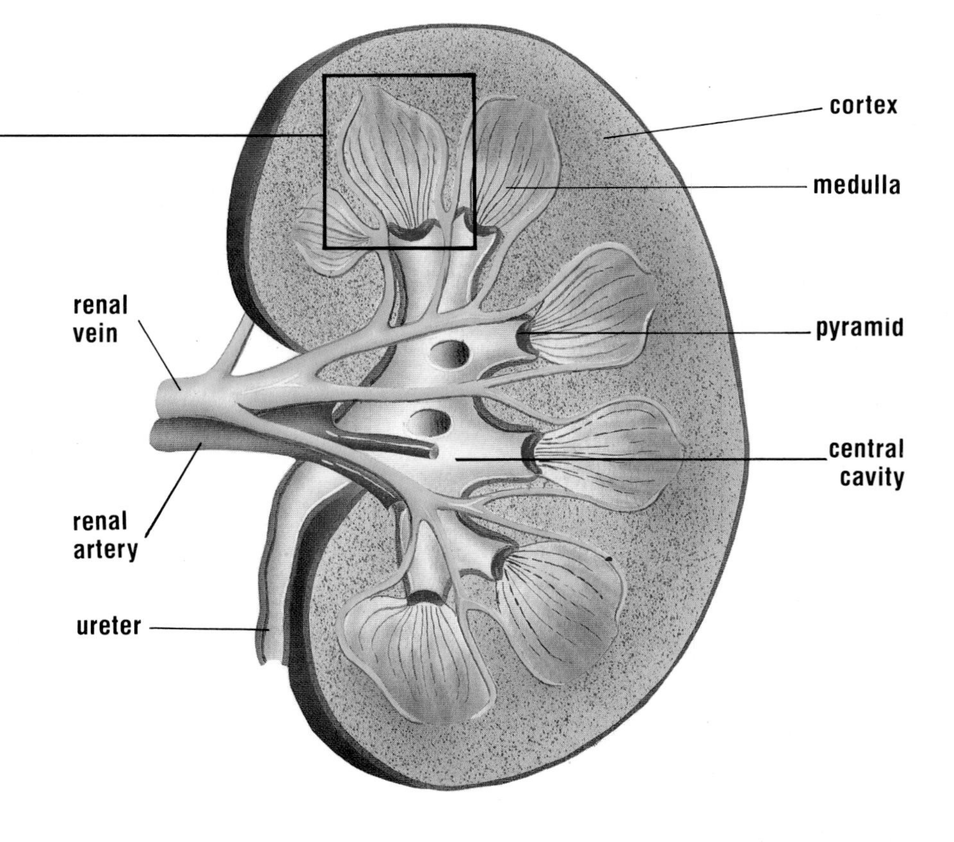

BLOOD

white blood cell

wall of blood vessel

red blood cell

phagocyte

The bloodstream is the most important transport system in the body, and each part of the blood has a particular job to do. But first, what are the different components that make up the 8 pints in an average adult?

Most of the blood is liquid — what we call plasma — and only 45 per cent consists of solid parts. Plasma is 90 per cent water; the remaining 10 per cent is made up of dissolved food and waste with some plasma proteins and hormones. The solid part consists of various cells: the button-shaped red blood cells, the much larger, irregularly shaped white blood cells, and tiny platelets.

Below: The people in the top row can safely donate blood to the people in the bottom row the arrows link them with. Group O blood can be given to anyone. People with Group AB blood can safely receive blood from anyone.

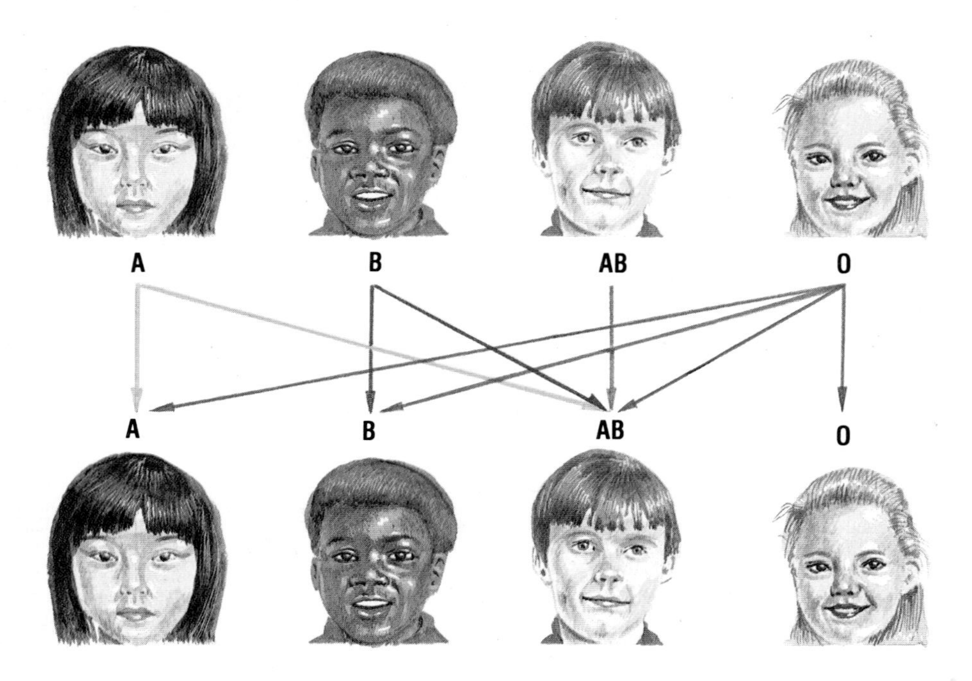

RED BLOOD CELLS

The most common type of cell or corpuscle is the red blood cell. There are approximately 600 red blood cells to every one white cell, which explains why blood is red. The red cells contain a red pigment, hemoglobin, and this chemical carries oxygen molecules around the body to help it do all the jobs necessary to keep us alive.

Red cells are unusual in not having a nucleus. They last for about four months and then are broken down in either the liver or spleen. Red blood cells are being made all the time and we replace about 2.5 million every second!

On the surface of red cells are special chemicals which decide our blood type, or group. Most people belong to one of the four main groups: A, B, AB, or O. Only group O people have none of these chemicals and so their blood is the only type which does not upset other groups' blood. Only they can donate blood to all other groups.

The job of the white corpuscles is explained on pages 56-57.

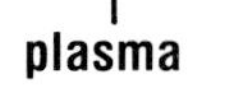

Above: A blood vessel containing red and white blood cells. Phagocytes and lymphocytes are types of white cell which deal with germs in different ways (see page 57).

Below: The network of arteries and veins which transports blood around the body. The arteries (shown in red) carry oxygen from the lungs to where it is needed. The veins (shown in blue) take carbon dioxide and other waste back to the heart and lungs.

PLASMA

Plasma is also used as an important transporter, because it carries material such as glucose and amino acids to all parts of the body. At the same time it also takes waste materials, such as carbon dioxide, away from the active tissues to the lungs to be breathed out. Another of the plasma's jobs is the transport of hormones around the body.

One other important job the blood performs is clotting. If it did not clot when we cut ourselves, valuable body fluids would be lost and germs would enter. Platelets and some of the proteins in plasma carry out this task.

Left: Blood donors are always needed so that hospitals can build up a store of blood of different groups. People who need blood following an accident or operation can then be given a transfusion to save their lives.

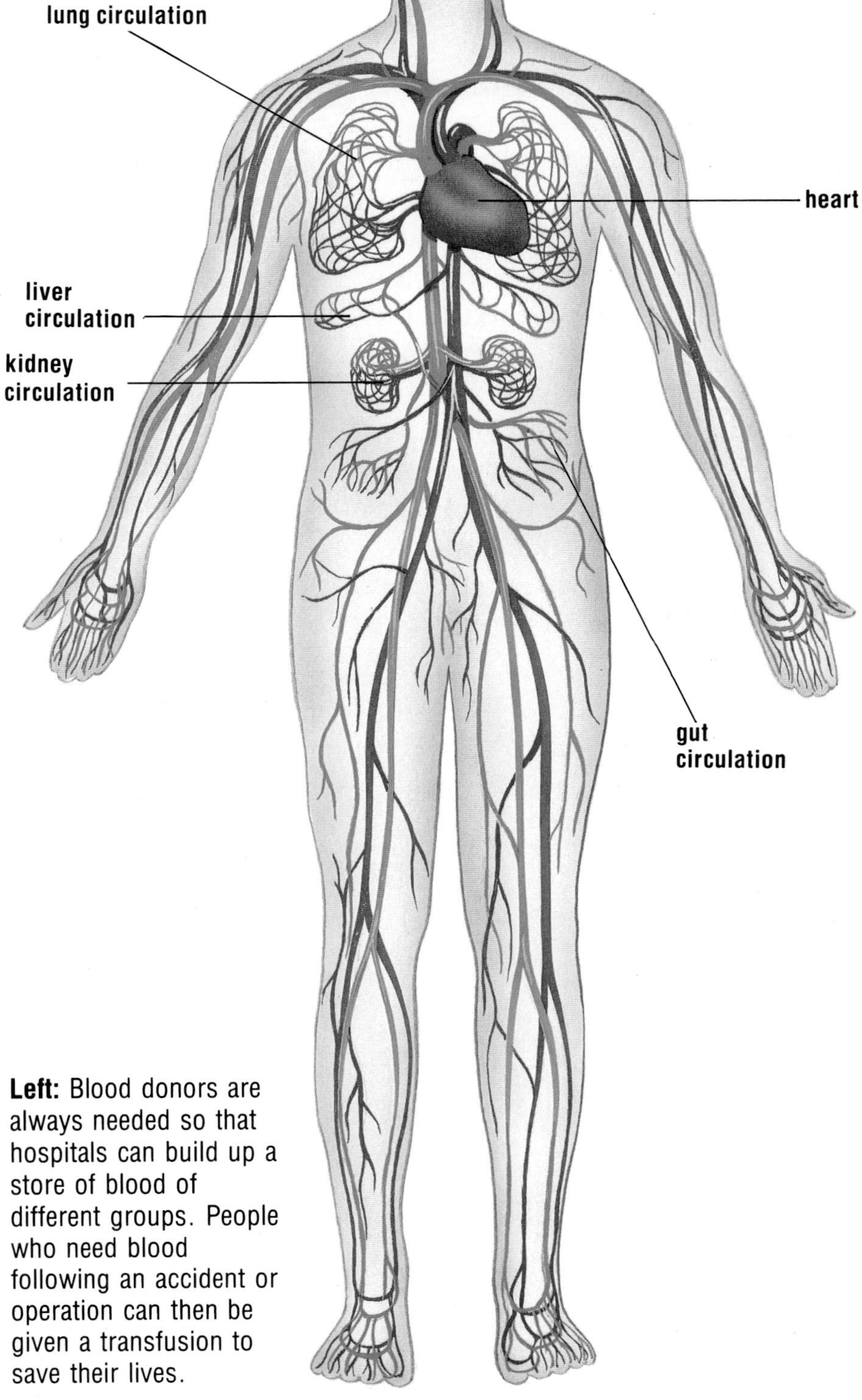

FIGHTING DISEASE

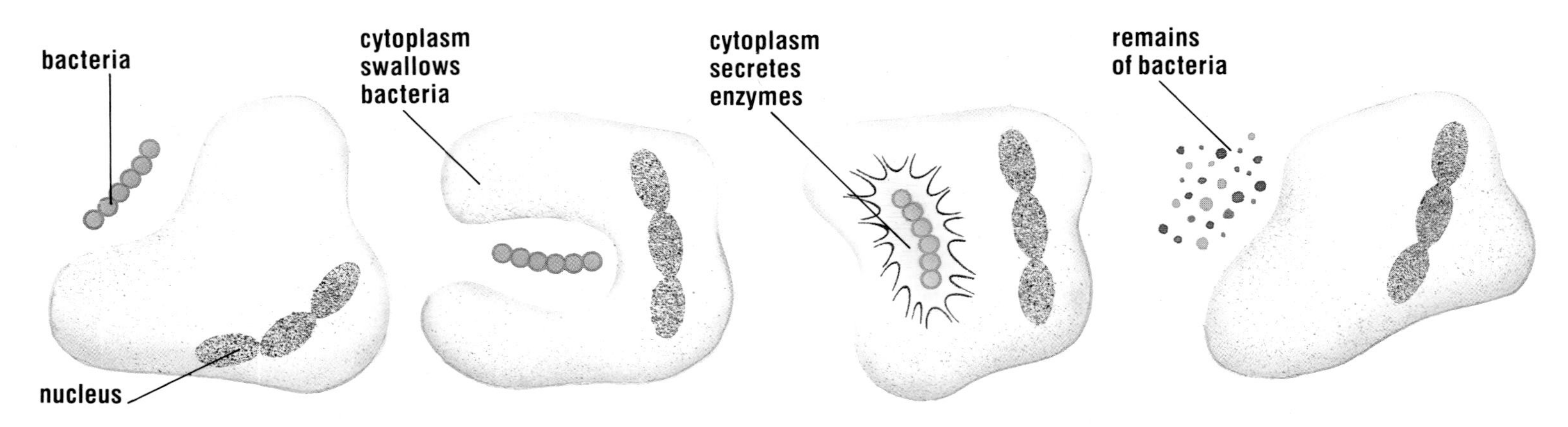

Above: A white corpuscle fighting bacteria which have entered the body. First it swallows them, then destroys them using enzymes. A harmless remainder is left.

Another vital job of the blood system is the fight against disease. Clotting of the blood prevents some germs from entering an open wound, but once germs have entered, by any route, the body has to defend itself.

The white corpuscles, or leucocytes, are the fighters. They are much larger than the red blood cells but there are far fewer of them. In 1 cu mm of blood there could be 5 million red blood cells, but only 6,000 white corpuscles. White cells also have a nucleus, unlike red blood cells, although this varies in shape depending on what type of white corpuscle it is.

Some white cells are capable of movement and creep along the inside of blood vessels by changing their shape as they go. Because they can change shape very easily, they can squeeze between the cells of the vessels' walls. Once out of the blood vessels, they move in the surrounding tissue and can swallow any germs they meet. This ability to eat organisms such as bacteria is very useful and can also take place in the bloodstream. Corpuscles that swallow germs are called phagocytes and they are being formed in the bone marrow all the time because their life span is very short — only a few days.

Below left: Children are immunized against certain diseases by the injection of dead or weakened forms of the disease-causing bacteria. Antibodies then form which could later fight the real disease.

Below: Phagocytes live for only a few days, but new ones are constantly being made in the marrow of some of our bones.

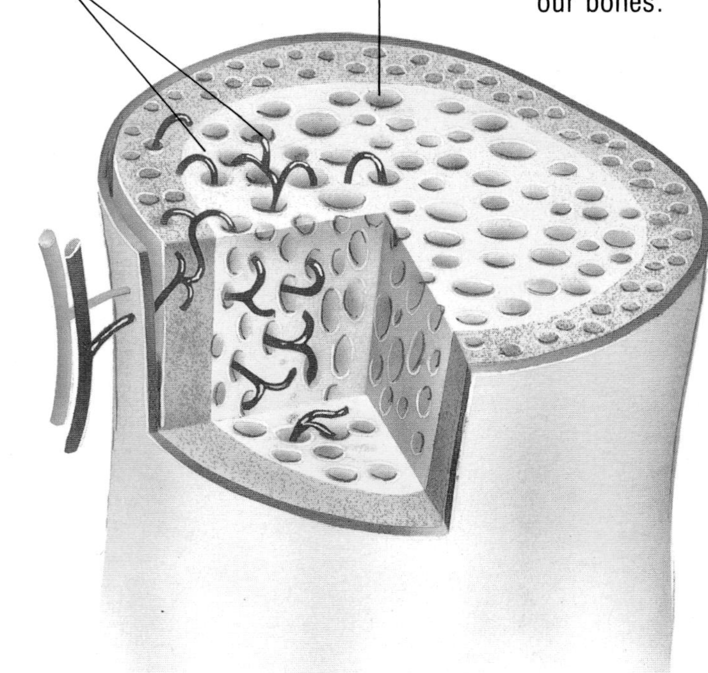

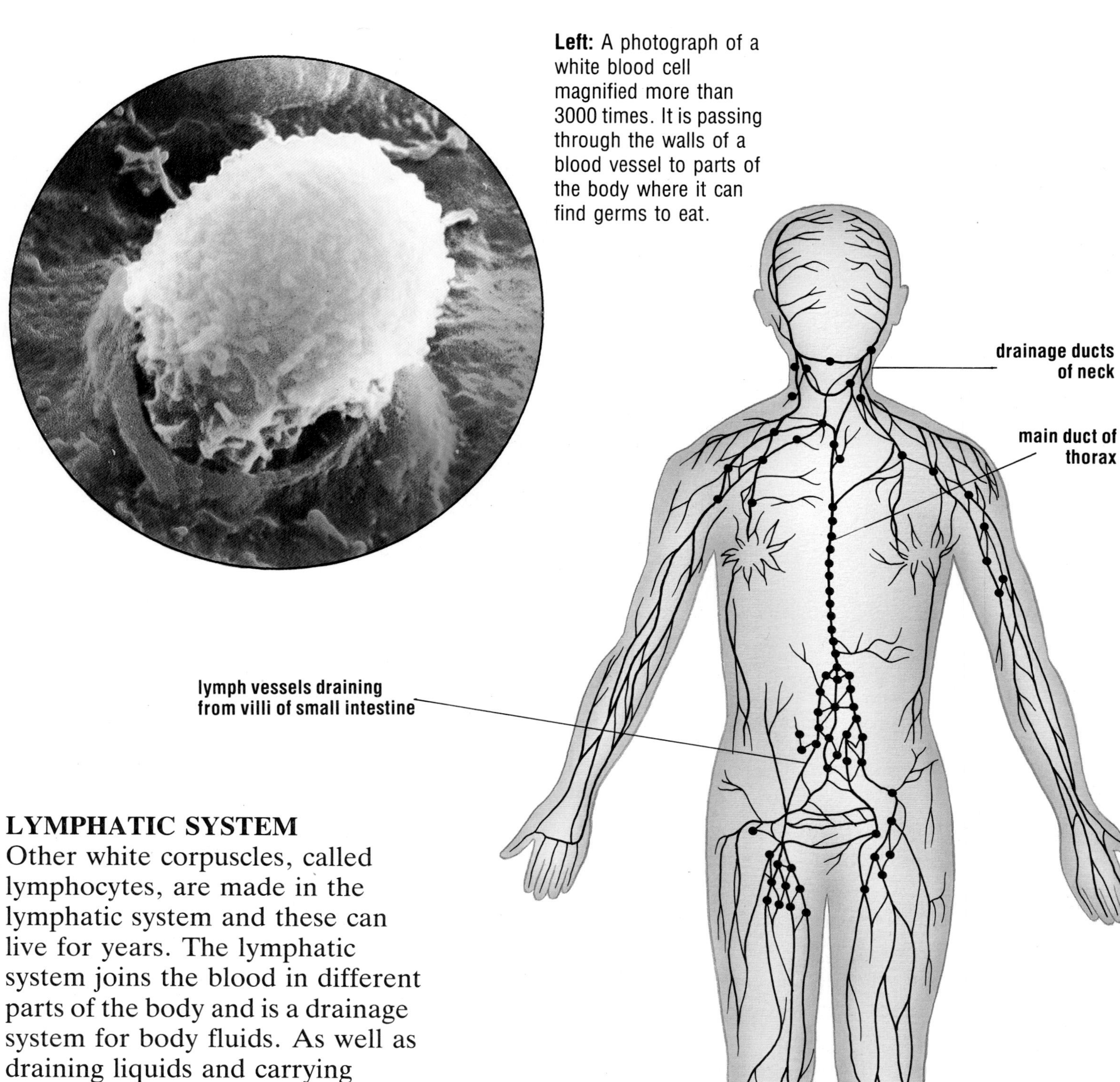

Left: A photograph of a white blood cell magnified more than 3000 times. It is passing through the walls of a blood vessel to parts of the body where it can find germs to eat.

Right: The lymphatic system of the human body. The dots indicate the positions of the lymph nodes.

LYMPHATIC SYSTEM

Other white corpuscles, called lymphocytes, are made in the lymphatic system and these can live for years. The lymphatic system joins the blood in different parts of the body and is a drainage system for body fluids. As well as draining liquids and carrying digested fats, the lymphatic system has the further task of producing lymphocytes in special swellings known as lymph nodes.

The lymph nodes act as filters, cleaning the lymph before it joins the blood system. Alien materials and germs are eaten by phagocytes or dealt with by the lymphocytes, which produce special chemicals called antibodies. These kill the germs or destroy their poisons.

THE HEART

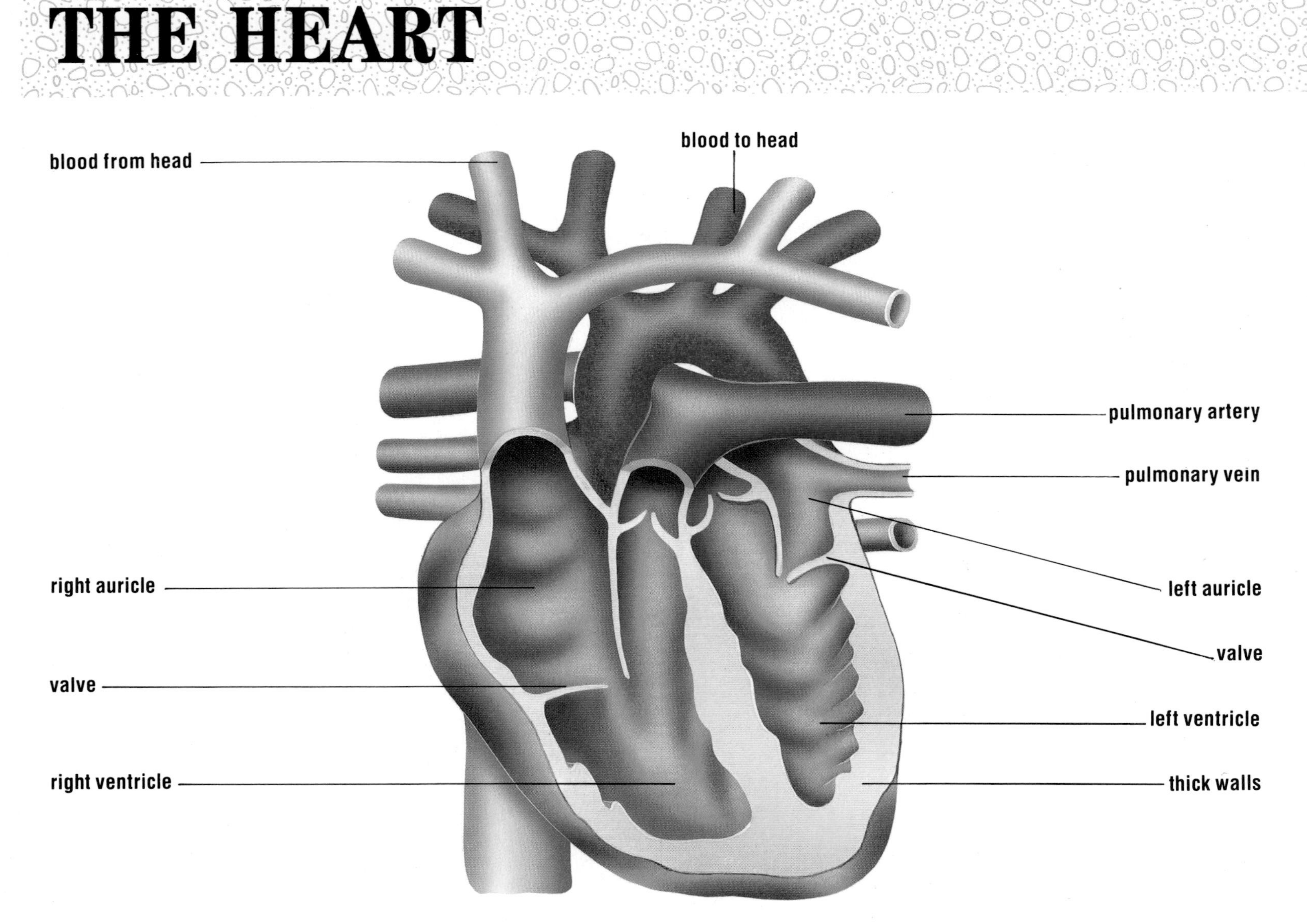

If the blood is to do its work, it has to be pumped around the body and this is why we have a heart. The human heart, like that of any mammal, has four "rooms," or chambers, and is situated between the lungs. It acts as a double pump, the right side receiving blood from which oxygen has been removed as it flows around the body and sending it to the lungs, and the left side receiving oxygenated blood from the lungs and sending it around the body.

The upper chambers of the heart are called auricles or atria. They receive the returning blood and then contract, forcing it into the ventricles below. Ventricles have much thicker muscles as they have to push blood a lot farther. The left ventricle, for example, must propel blood all around the body before returning it to the right auricle.

To keep the blood moving and prevent any backflow, there are one-way valves at the openings into the auricles, ventricles, and

Above: The human heart is a complex structure with four chambers, two thin-walled auricles and two thick-walled ventricles. Its function is to pump blood all around the body.

Below: The trace of an electrocardiogram, a machine which records the heart's electrical activity. Doctors use this machine to check if a patient's heart is healthy.

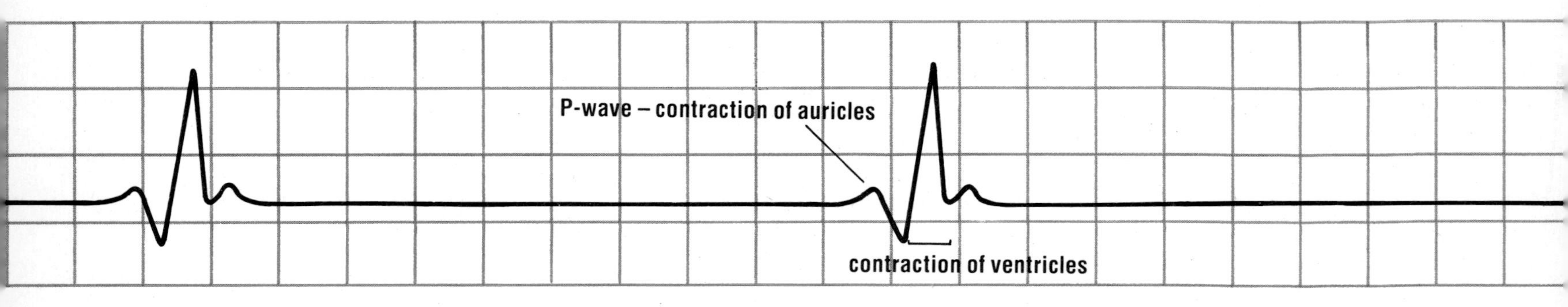

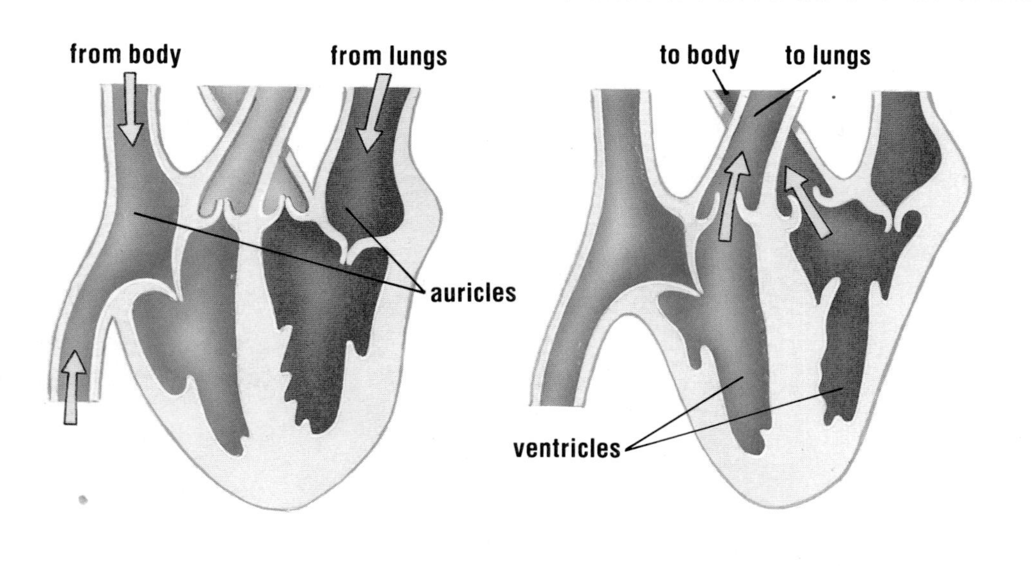

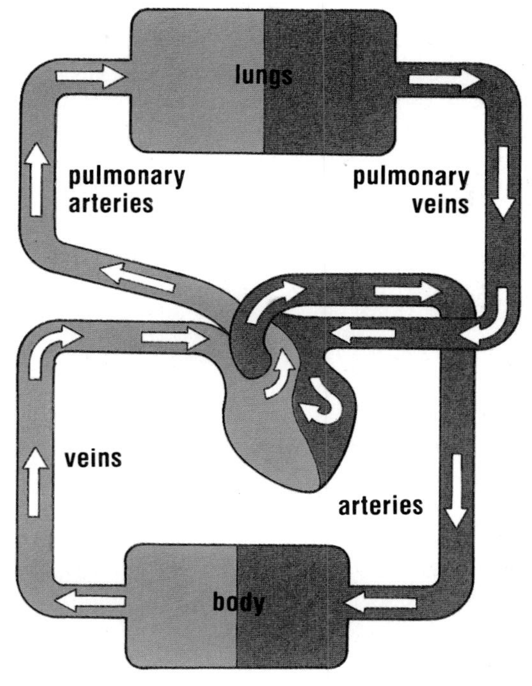

large arteries leaving the ventricles. The valves between the auricles and ventricles are the strongest as they have tough cords which prevent the flaps of the valves being turned inside out.

The two auricles contract at the same time, and then the two ventricles, so that blood leaves the heart in spurts along the major arteries, and the blood vessels have to be able to enlarge slightly as blood flows along them. You can test this pumping action indirectly if you gently feel the pulse on the underside of your wrist.

The normal resting pulse is about 70 beats a minute, and it rises when we are active because the heart must beat faster to pump blood more quickly around the body to active muscles. The hormone adrenaline can also cause the heartbeat to speed up, when we are frightened or excited.

Above: The left and right halves of the heart have different jobs. The right half receives deoxygenated blood from the rest of the body and returns it to the lungs. The left half receives oxygenated blood from the lungs and sends it around the body.

Above: How blood circulates around the body. The blood always flows in the direction shown. It takes less than one minute for a complete circuit to be made.

Left: Our pulse is caused by the heart pushing blood along the arteries. The beat can be felt by placing the tips of your fingers very gently on the underside of the wrist. The thumb should not be used for this as it has a pulse itself.

THE LUNGS

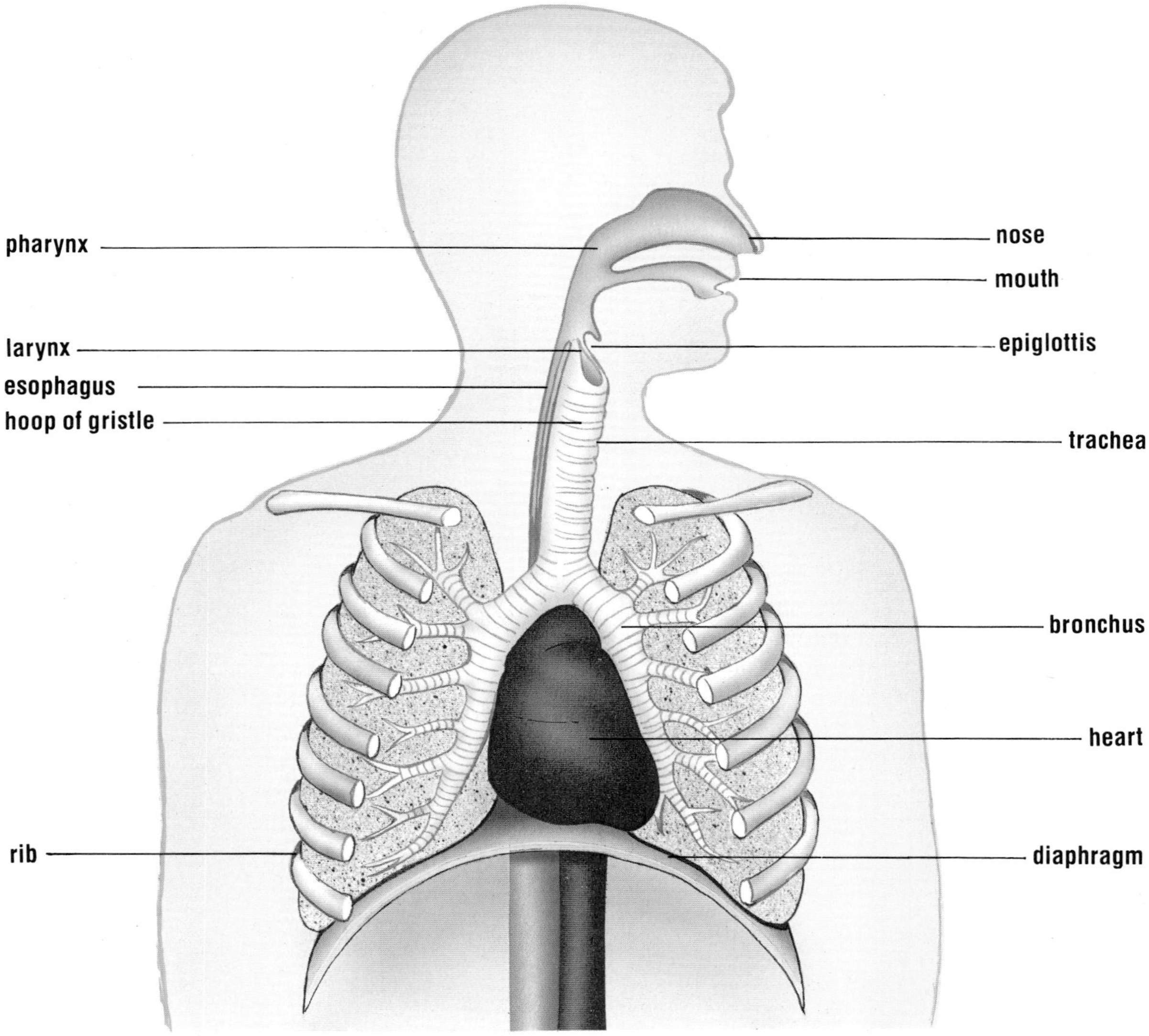

Right: Although people are not designed to spend long periods under water, modern diving equipment makes this possible. Tanks of compressed air carried on the back enable the lungs to keep taking in the oxygen they need.

Breathing is essential for human life. From the air we breathe we take oxygen, which we need if our bodies are to turn the food in the blood into raw energy.

Like all mammals, humans need lungs to take oxygen from the air they breathe in. The lungs look like two large, pink-gray spongy bags. They are situated in the chest and protected by the ribs. It is through the linings of the lungs that the oxygen passes. Molecules of gas move slowly through the lining cells, dissolved in the liquid on the inner surface of the lungs.

To force air to move in and out of the lungs the chest has to expand and contract. Powerful muscles move the ribs both upward and outward and another muscle just below the lungs, called the diaphragm, flattens out to expand the chest area. When the chest is expanded, the air pressure inside the lungs is low and the air pressure outside the body is high enough to push air in. When the chest contracts, the pressure inside the body forces air out.

When we breathe in, air enters through either the mouth or the nose. It is better to breathe in through the nose because the air is then warmed, moistened, and has dust removed from it. Go to sleep with your mouth open and it will feel very unpleasant and dry when you wake up because air has been passing back and forth in it.

Above: The parts of the body we use for breathing are called the respiratory system. Air passes into the nose or mouth, down the trachea, through the bronchi and into the lungs. The lungs fit tightly inside the ribs, which protect them. Movement of the rib muscles and diaphragm allow us to breathe by moving the lungs in and out.

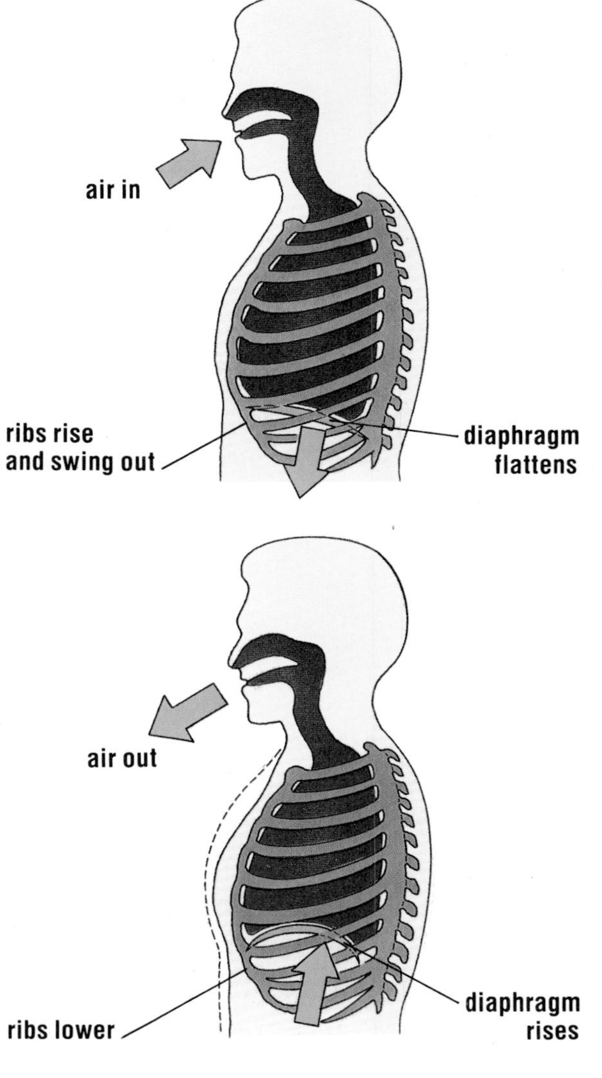

Above: To breathe in, the diaphragm flattens and the ribs rise and swing out. To breathe out, the diaphragm rises and the ribs lower.

As we breathe in, the epiglottis, or flap at the top of the windpipe, opens to let the air rush down. Normally the epiglottis is closed to prevent saliva and food from entering the windpipe, or trachea.

The trachea eventually reaches a fork, where it divides into two branches, or bronchi, one leading to the right lung, the other to the left. Both the trachea and the bronchi have rings of gristle around them to help keep them open at all times. Each bronchus divides again and again, eventually forming many fine tubes which end in clusters of air sacs where the oxygen is absorbed and the waste carbon dioxide gas we breathe out seeps back into the lungs.

Below: Athletes burn up a lot of energy. They use oxygen to produce this energy, so often pant after a race to take more oxygen in.

ABSORBING OXYGEN

Once the oxygen in the air reaches the air sacs, it has to spread through the lining of moisture and the thin walls of the lungs so that it can be carried away in the blood. At the same time waste carbon dioxide passes from the watery blood plasma, through the walls of the air sacs, and into the cavities to be breathed out.

When we are very active our muscles work hard and produce extra waste carbon dioxide. This carbon dioxide dissolves in the blood plasma and is carried around the body. There are special sensory cells in the walls of the blood vessels of the neck which check how much carbon dioxide there is in our blood. If there is more than usual, then the sensory cells send messages to the brain. The brain interprets the incoming signals and tells the rib muscles and diaphragm to contract faster, and so we breathe faster to get rid of the extra carbon dioxide.

Air passing through the nose is filtered for dust, but the job is finished by two other processes. Lining the windpipe, or trachea, are thousands of tiny hair-like outgrowths called cilia. These contract gently all the time, pushing the slimy moist liquid on the lining of the trachea upward to the mouth and nose. This sticky layer traps dust so particles do not pass downward. The second method of checking the entry of dust is a type of cell which lines the air sacs and can swallow dirt particles.

DANGERS OF SMOKING

Both these valuable cleansing mechanisms are seriously damaged by smoking. Smoke taken into the body paralyzes the cilia so that slime and dust slowly slide down the trachea to the lungs, causing coughs and possible bacterial infection. The amount of dust in the smoke also proves too much for the cells lining the air sacs and therefore more dirt accumulates in the lungs. As time passes, other grave diseases may result because of the dangerous chemicals in the smoke (see page 72).

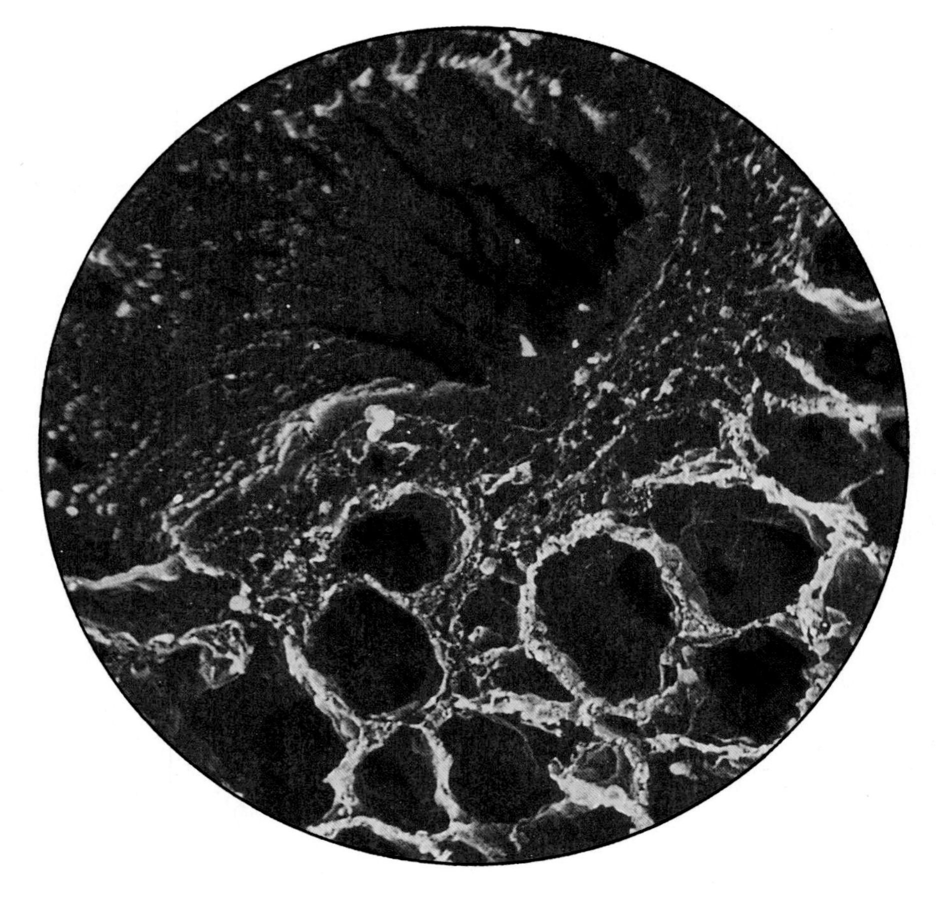

Above: A greatly magnified photograph of a balloon-shaped air sac or alveolus.

Below: A magnified photograph of the cilia which line the trachea and bronchi. They constantly beat upward to drive dust out.

Right: Inside, our lungs look like this. The bronchi gradually divide into smaller and smaller branches to carry air deep into the lungs, ending in tiny air sacs or alveoli. Each air sac is surrounded by blood vessels. Oxygen passes through the thin walls of the sacs into the blood and carbon dioxide passes back through these walls to be breathed out.

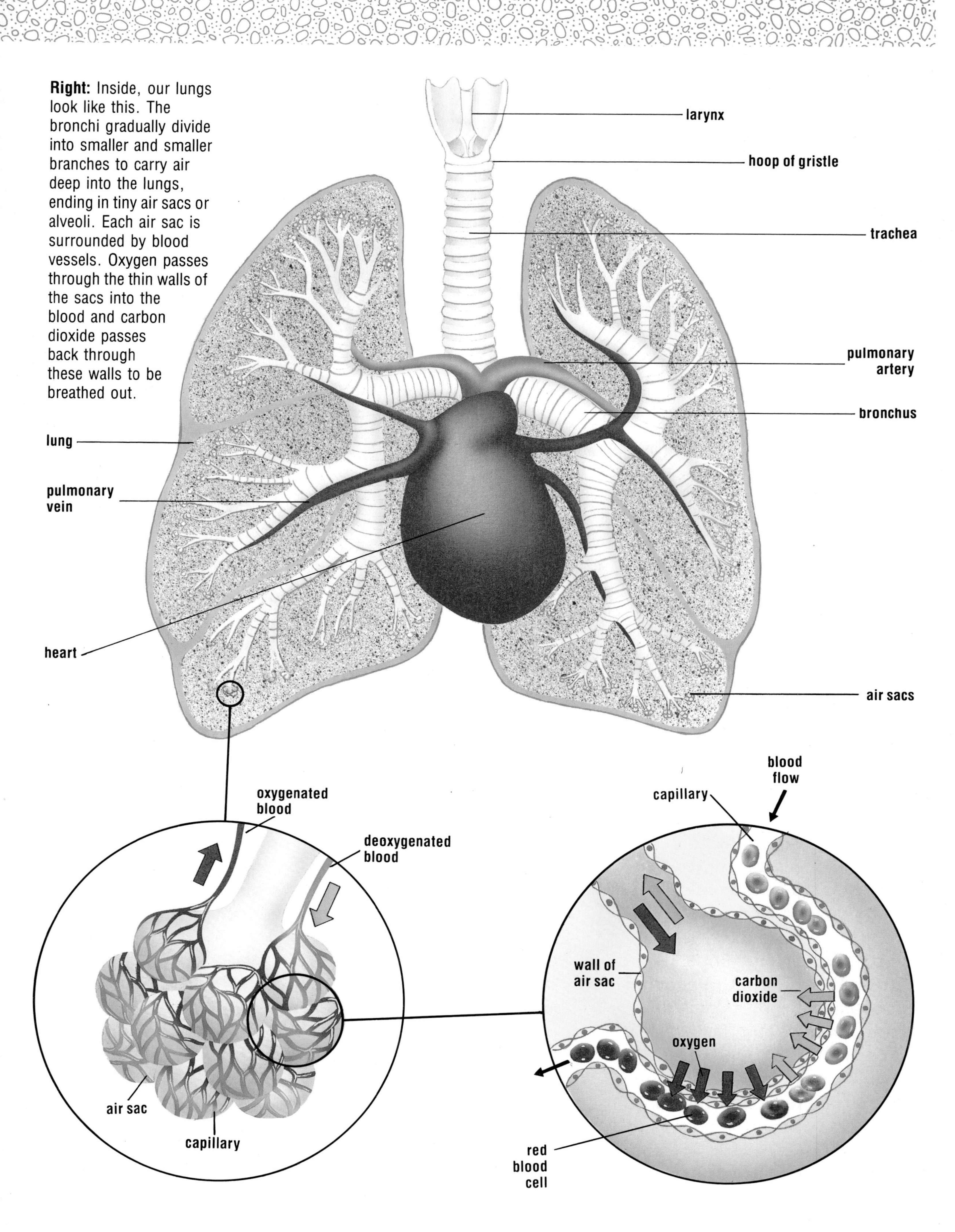

REPRODUCTION

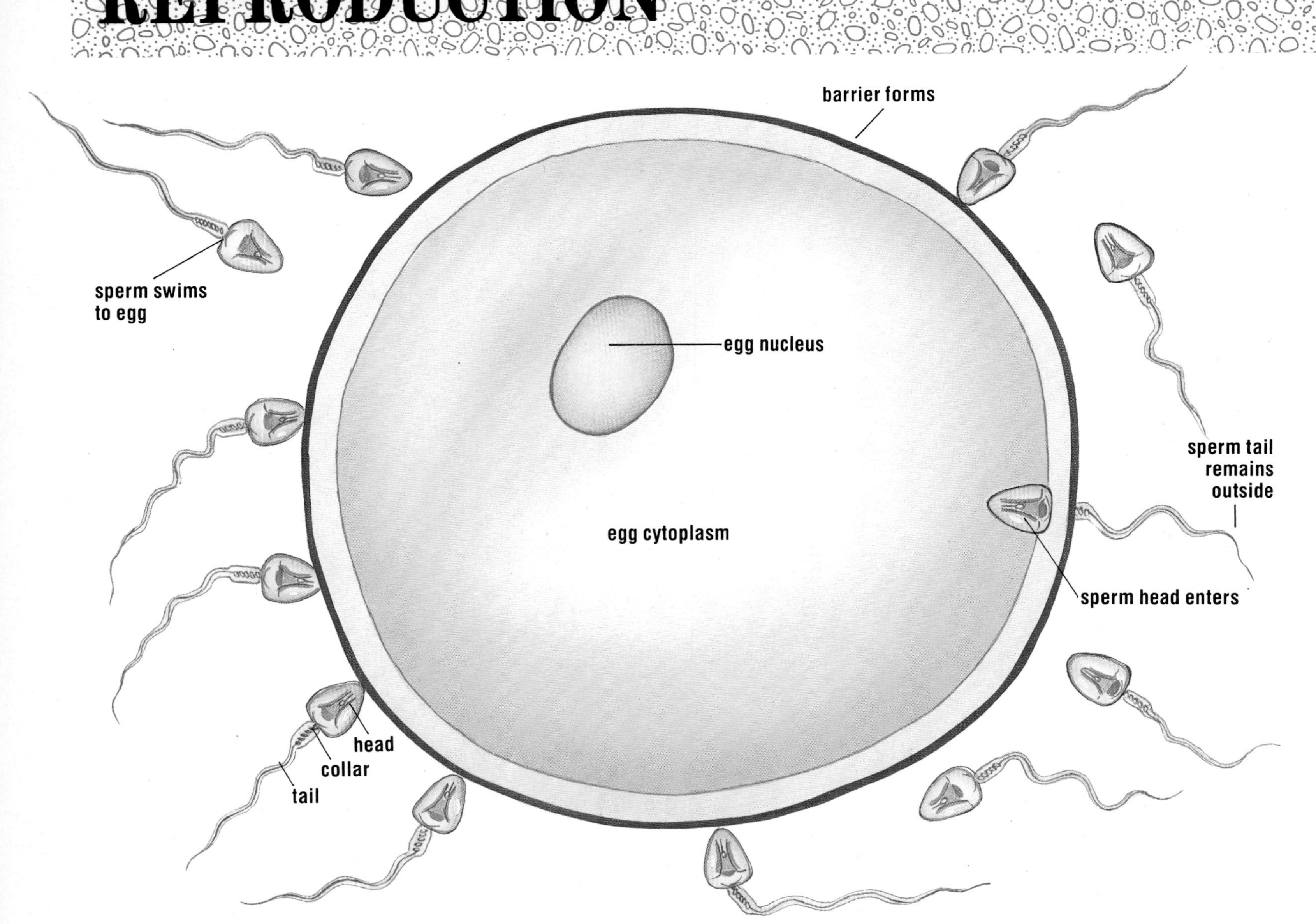

To make a new human life, a sperm cell from a man has to penetrate an egg cell from a woman in a process called fertilization. It is only after puberty that males produce sperm cells and females make egg cells. This change, like many of the major differences between men and women, is caused by special sex hormones. In girls, the hormones also result in the female body's developing different proportions to that of the male. Breasts develop, hips widen, and the outlines of the body become a series of smooth curves due to a layer of fat. In boys, the body develops a heavier frame with larger muscles. Young men also develop other features which make them different from boys, such as the growth of hair on the face and deeper voices.

The female monthly period, or menstrual cycle, starts at different times in different people, usually between the ages of 11 and 15. The production of an egg from one of the ovaries occurs midway through the cycle, and although the timing is erratic at first, it becomes more regular in the late teens. At the end of the cycle, if the egg is unfertilized, the uterus, or womb, sheds the lining prepared for the egg. This causes a flow of blood lasting about four to five days, and then the cycle begins again.

Males usually begin to produce sperm in the testes between the ages of 11 and 15. Production goes on all the time and the sperm are

Above: A sperm entering an egg to fertilize it. The sperm's tail remains outside. Once the sperm has entered, a barrier forms to prevent other sperm from going in.

Above right: A photograph of a fertilized human egg. The egg now contains all the information needed for the development of a child.

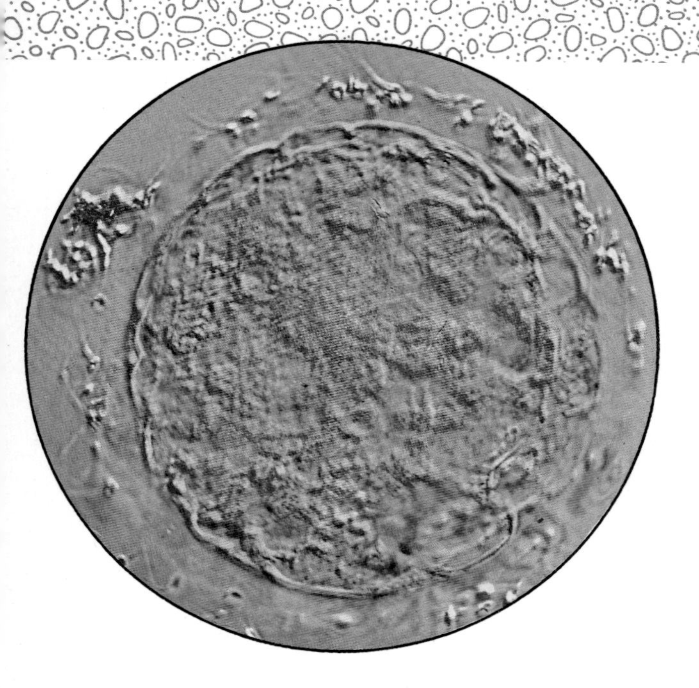

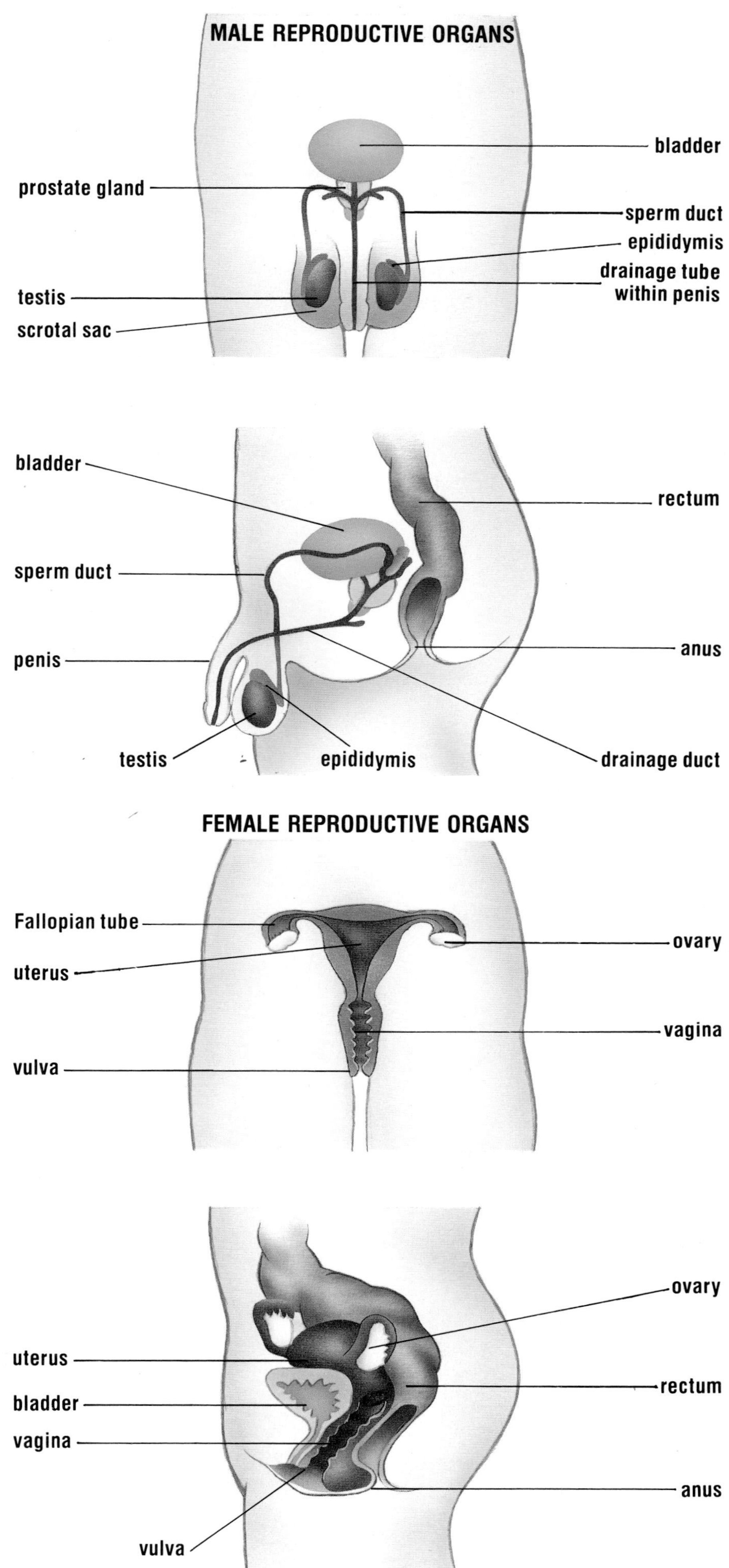

stored in a series of coiled tubes called the epididymis. If not used, the sperm eventually die and are absorbed by the body tissues.

Fertilization, or the joining of the female egg and a sperm, takes place within the woman's body following sexual intercourse. The sperm swim up the woman's vagina into the womb, or uterus, and along the egg, or Fallopian, tubes. Movements of the walls of the vagina and uterus help them make this journey.

Sperm and egg join when the head of a sperm penetrates the outer layers of the egg and their nuclei fuse together. This fusion is the act of fertilization and is the moment when hereditary material from the father and mother meet. Fertilization can occur only in the upper part of the egg tube. The fertilized egg is the first cell of the next generation and the entire baby will develop from it.

PREGNANCY AND BIRTH

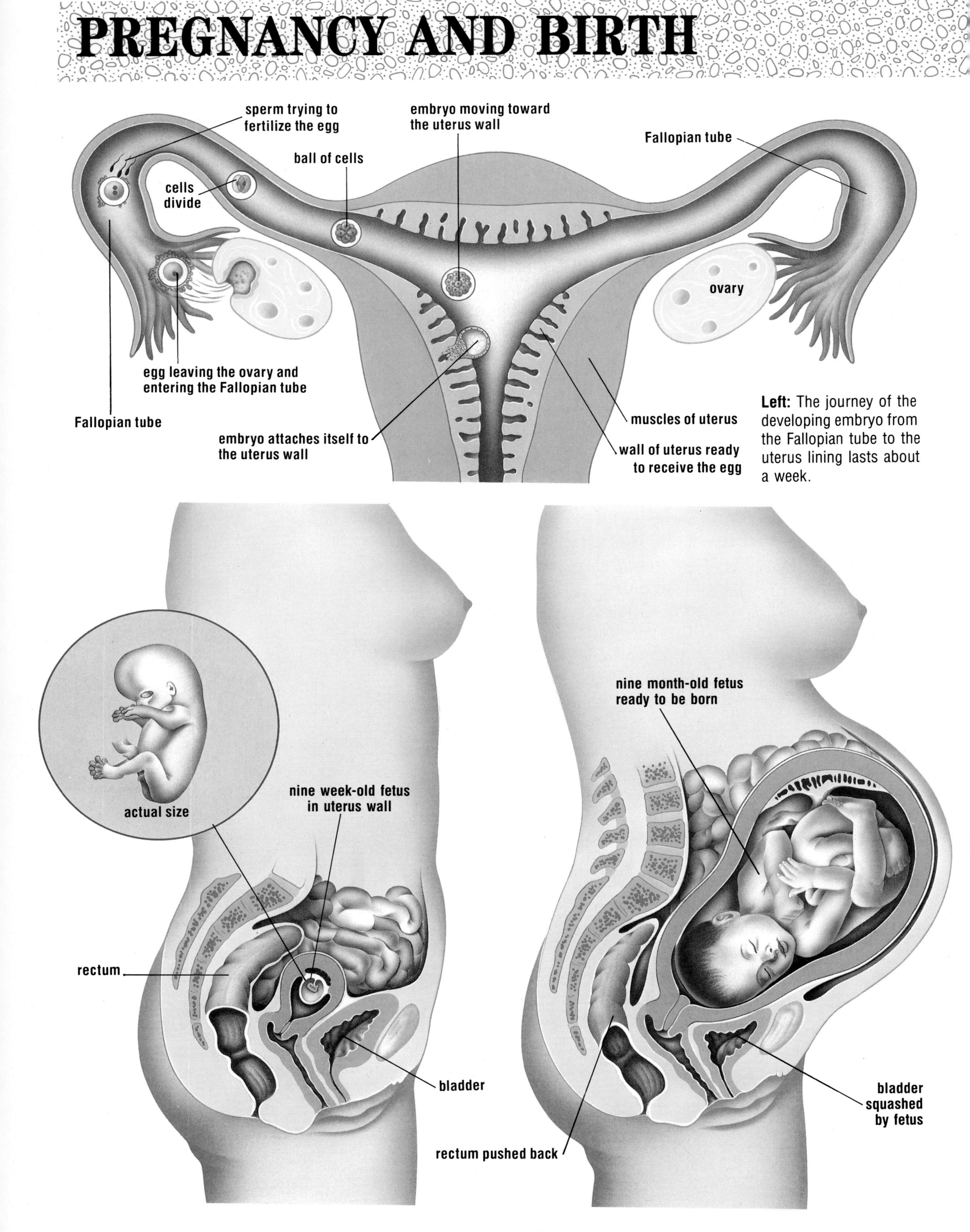

Left: The journey of the developing embryo from the Fallopian tube to the uterus lining lasts about a week.

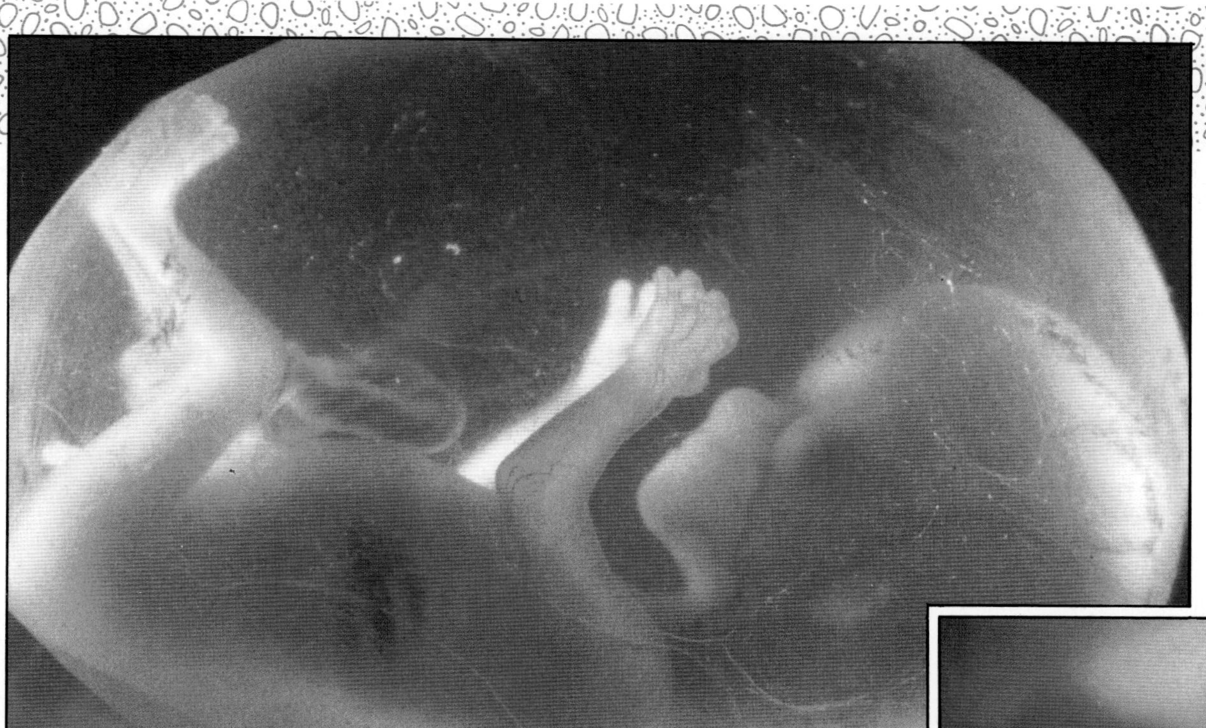

Below: At five months the fetus is well developed and it is possible to tell whether it is a boy or girl using an ultrasound scan. The head has a covering of fine hair, and eyelids and nostrils can be clearly seen.

Above: A photograph of a 14 week-old fetus. It lies in a pool of liquid which supports it and keeps it at a temperature of 98.6 °F. Food from the mother arrives via the placenta. At this stage the fetus can kick, frown and even move its fingers.

Far left: The embryo develops within the uterus and by nine weeks the main parts of the body have been made. However the mother will not look obviously pregnant.

Left: At nine months the fetus is ready to be born. The mother is visibly larger and her bladder and rectum are squashed.

After fertilization, the single new cell moves along the egg tube and down into the uterus. It takes approximately a week to reach the uterus and as it travels it divides again and again to form a hollow ball of cells called an embryo. Once in the uterus, it sinks into the spongy walls and "roots" itself. Here the mother's body will supply the young embryo with food and oxygen, both of which are absorbed into the hollow ball.

As time passes, the embryo grows larger and starts to change shape. It takes about nine months for a baby to develop fully in the uterus, but it assumes a human shape after only two months. From this stage onward it is called a fetus.

If it is to grow all the different tissues and organs of the human body, the embryo must have very good supply and waste-disposal systems. The placenta is the structure through which food and oxygen pass to the blood supply of the embryo along the umbilical cord, while waste passes the other way, from the embryo to the mother's blood.

But it is not only useful raw materials that can pass from the mother to her unborn child. It is important, therefore, that pregnant mothers do not take medicines unless their doctors believe it to be absolutely necessary. Drugs such as alcohol and tobacco can also upset the unborn child. Some viruses, such as German measles, can cause abnormal development in the fetus, too.

After nine months the baby is at last ready to face life outside its mother's body, and it is forced out by strong contractions of the muscles in the walls of the uterus. The umbilical cord is cut to free the baby from its mother. Then the uterus contracts to force out the placenta, now called the afterbirth.

HEREDITY

Most of us look something like our parents. We have the same color eyes, perhaps, or the same color hair. This is because some features, such as eye and hair color, are inherited.

We have special units called genes in the nuclei of all our cells, which carry instructions about how to make an exact copy of our whole body. These genes are made from a special chemical called DNA, which forms pairs of structures known as chromosomes. One member of each pair comes from our mother, the other comes from our father, and we inherit them at the moment our father's sperm cell fertilizes our mother's egg cell.

Some genes are powerful in their effect and they force us to show the feature they carry, whether their partner genes are the same or not. Powerful genes like this are called dominant. Weaker genes are called recessive and we show the feature they carry only if both members of the pair are recessive. A "mixed" pair — with one dominant gene and the other recessive — results in our showing the feature of the dominant gene.

Many human features are the result of the combined activities of many genes, but there are a few that are controlled by only a single pair. Examples of these single-pair features include blood groups – A and B are dominant, group O is recessive — and eye color, where brown is the dominant gene and blue is recessive.

One simple feature that is very easy to check is the ability to roll the tongue into a tube. Anyone who can has at least one dominant gene for "rolling" in the pair of genes used to control this feature. "Non-rolling" is recessive and therefore someone who cannot roll

Below: Whether our ear lobes are attached to or free from the face is controlled by a single pair of genes. Free lobes are dominant and attached lobes recessive. The grandfather of the family has attached lobes, but his wife's are free. Through the dominant gene, their son has inherited free lobes, but he still carries a "hidden" recessive gene for attached lobes. His wife must also have a hidden recessive gene, as her daughter has attached lobes. She has inherited the recessive gene from both parents. Eye color is also determined by genes (see caption page 69).

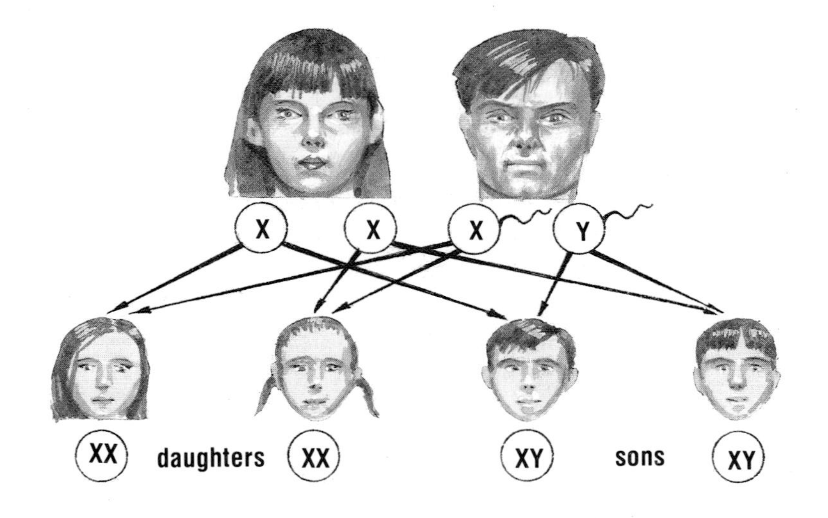

Left: All females have two X chromosomes and all males have one X and one Y chromosome. Half a man's sperm have a single X chromosome and half a single Y chromosome, but every egg contains an X chromosome. If an X sperm joins with an egg at fertilization, a baby girl will be born. If a Y sperm joins the egg, the baby will be a boy.

the tongue must have inherited non-rolling from both parents.

Although features like eye color and tongue rolling are always shown, there are times when certain genes are not expressed, even though they are dominant. For example, if a person who had the genes for growing very tall ate a very poor diet when young, he or she might remain small and never reach that potential.

Below: Non-identical twins are the result of two sperm fertilizing two different eggs. Identical twins develop when a single fertilized embryo divides.

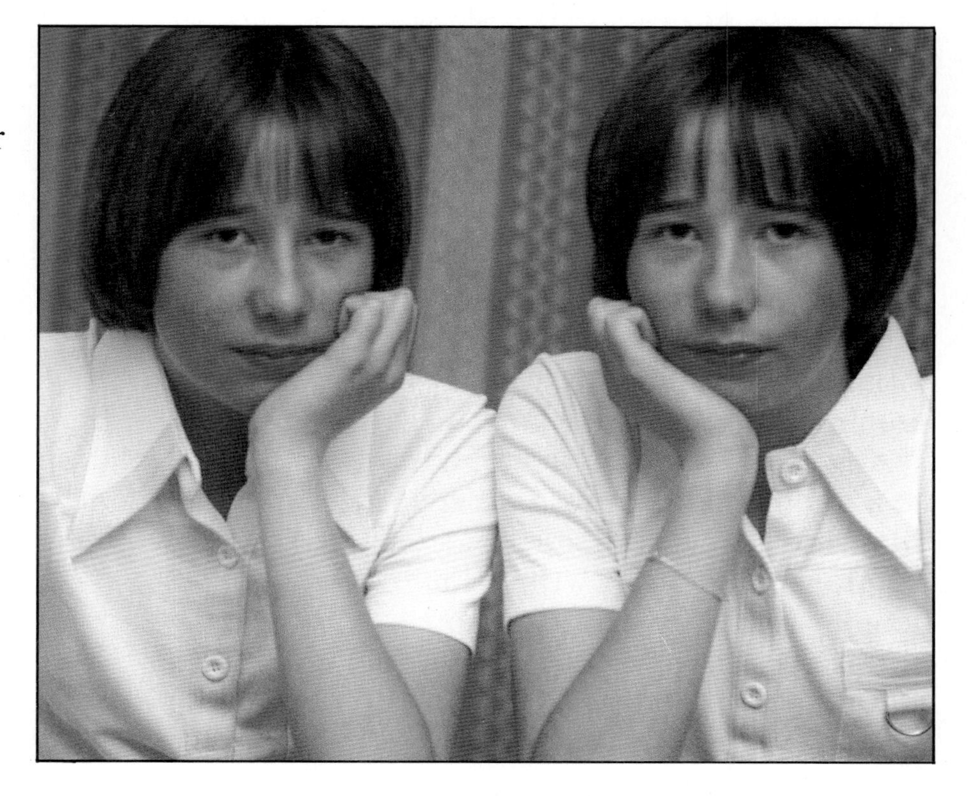

Left: Like ear shape (see caption page 68) eye color is also controlled by a single pair of genes. Brown eyes are dominant and blue recessive. The grandfather must have had a hidden recessive gene for blue eyes which paired with the grandmother's blue eye gene to produce a blue-eyed son. This recessive gene pairing has continued through the generations.

GROWING UP AND OLD AGE

We all pass through a series of stages as we grow up. The most important is the pre-school stage between birth and five years old. It is in this early period of our lives that we learn how to use our bodies, how to speak, and how to live happily with others.

Our physical development in the first two years helps us learn to walk, hold things, and generally operate our muscles. In one experiment in the USA, a family raised a newborn ape alongside their child. Physically the ape learned faster than the human being. However, the child quickly caught up and passed the ape at about 18 months to two years once it began to speak. It is because we can communicate by talking that, though relatively weak, we are such dominant animals.

All records of growth can only be a rough guide because we are all different, but it is possible to detect different rates of growth in the height of boys and girls. Also, different parts of us grow at very different rates. Our heads, for example, are always relatively large when young because of the enormous development of the brain. The rest of our bodies only slowly grow into proportion.

During adolescence our bodies undergo great changes, both chemically and physically, as we become more adult. The chemical changes are started by the pituitary gland in the brain and then controlled by the sex hormones. By the age of 16 to 19 years most growth has finished and the body must now maintain its rate of cell division to repair and replace cells for many years to come.

Above: At age 9, the head is still large in proportion to the body, but by 13 this is no longer so. Instead, following puberty, both boys and girls start to grow faster and in the girl breasts begin to develop. By the age of 17, most growth has finished and the body has its adult shape.

Left: During their adolescence, young people not only develop physically but change in their attitudes toward each other. Friendships between the sexes become more important.

13 years old

17 years old

GROWING OLD

Ageing comes to us all. The skin becomes wrinkled, the hair turns gray, and in men may fall out. The skeleton loses calcium salts and becomes more fragile, and the joints get stiffer as less exercise is taken. Lack of exercise reduces the muscle tissue, too, and we also feel the cold more. Finally, as the nervous system deteriorates, hearing gets worse and the lenses of the eyes become stiffer.

Right: The child has smooth skin and thick hair, but the man has lost these with age.

ENVIRONMENT

The world around us affects us all. When conditions are bad, our bodies do not work well, and when our bodies are not working well we do not feel good. There are many reasons for concern about the world around us and how it affects our health.

Air pollution is a major problem. It can be the result of industry, for example, sulfur fumes that contribute to acid rain or dust from brickworks and steelworks. Both sulfur and dust can cause damage to our lungs. Many people in Eastern Europe, for example, have damaged lungs, livers, and eyes due to the air pollution from very inefficient factories.

Air pollution can also be caused by trucks and cars which produce various exhaust fumes. The major health hazards from vehicles are carbon monoxide, which harms red blood cells, and lead, which can upset the development of young children. The use of lead-free gasoline helps reduce this problem.

Smoking is another air pollutant, and smoking in enclosed spaces can cause harm to the smoker and to others. Diseases known to result from smoking include lung cancer, bronchitis, and heart disease.

What we do with our waste materials can also affect our health. Our way of life in Western Europe and the USA produces a great deal of waste. It must be carefully treated if it is not to attract germ-spreading flies and rats or give off harmful gases. Household garbage is often buried, but this can have long-term dangers.

Factory farming is another threat to health. Over the last 20 years farmers in the USA and Europe have produced vast mountains of food from the land. But it has taken a great deal of fertilizers and pesticides to do it. Fertilizers have washed out into our rivers, causing the excessive growth of algae that produce poisons, while the pesticides that killed crop pests can also be passed up the food chain and may harm our health. Disease in factory-farmed animals is common too, and also affects us.

But not all the news is bad. Many people in Western Europe and the USA live in much better houses than their great-great-grandparents did 100 years ago. Clean drinking water and efficient sewage systems have eliminated many of the dreadful diseases of previous centuries, such as cholera and typhoid. But in parts of the world without clean water and safe sanitation, or following a natural disaster like an earthquake, such waterborne diseases are still a threat to life.

Above: The air in many of our cities has become very unpleasant to breathe because of pollution. This photograph shows smog, a mixture of fog and smoke, in Los Angeles, California, USA. Smog can be caused by the sun shining on exhaust smoke from cars.

Far right: Some of the damaging effects that air pollution and drinking alcohol may have on the whole body.

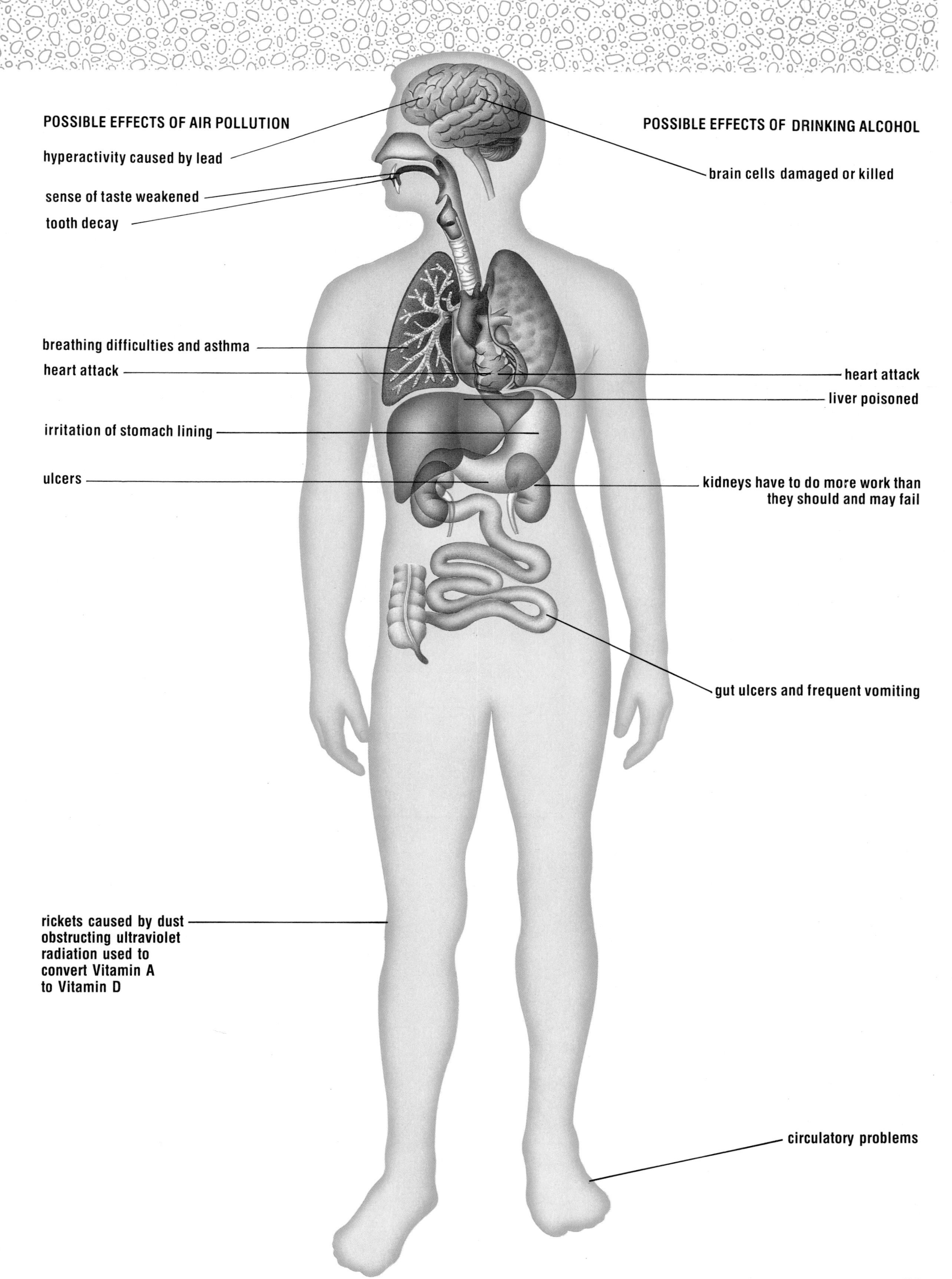
POSSIBLE EFFECTS OF AIR POLLUTION
hyperactivity caused by lead
sense of taste weakened
tooth decay
breathing difficulties and asthma
heart attack
irritation of stomach lining
ulcers
rickets caused by dust obstructing ultraviolet radiation used to convert Vitamin A to Vitamin D
POSSIBLE EFFECTS OF DRINKING ALCOHOL
brain cells damaged or killed
heart attack
liver poisoned
kidneys have to do more work than they should and may fail
gut ulcers and frequent vomiting
circulatory problems

DIET AND EXERCISE

At birth, many children are breast-fed, that is they are fed on their mother's milk. This contains all the nutrients they will require for the first few months of life, plus the antibodies they need to protect them against infection. An alternative is the various formulas for bottle-fed babies.

However, after a few months babies start to take solid food, and by the end of the first year they are eating a diet similar in content to what their parents eat. A balanced diet for adult and child contains seven major components. We need all these foods throughout our lives, but the amounts we need vary with our activity, age, and sex.

Carbohydrates provide energy for activity and also keep us warm. Fats act as an energy storehouse and also may contain certain vitamins, such as A, D, E, and K. Proteins enable the body to make cells and structures such as hair and nails, so they help us grow and repair tissue. Fiber keeps the gut working properly because it helps prevent constipation.

Above: The hamburger this boy is eating contains carbohydrate in the form of starch and protein in the meat. Both these things are vitally important in our diet, but on their own are not a balanced meal. The meat may also contain too much fat. The amount and type of food we need depends on our age, sex and activity level. A young, growing child will need more protein than an older person sitting in an office all day using up little energy.

WHICH EXERCISE IS BEST?

Exercise	Strength produced	Stamina produced	Agility produced	Energy used (kilojoules/min)
Walking to school	★	★★	★	10
Fast walking	★	★★★	★	20
Slow cycling	★	★★★	★	20
Jogging	★★	★★★★	★★	25
Swimming	★★★	★★★★	★★★	35
Fast cycling	★★★	★★★★★	★★	over 40
Disco dancing	★	★★★	★★★★	over 40

Left: The chart shows which of a variety of exercises scores highest for producing strength, stamina and agility.

Left: Cycling is a good form of exercise, helping us to stay fit and keep our heart and lungs in good condition.

Above: Swimming exercises many of our muscles without straining them. The water supports our weight so that there is not too much pressure on our skeleton. If we are able to visit an indoor pool, swimming is a healthy form of exercise we can do all year.

Vitamins keep all the complex chemistry of the body working. Minerals work with the vitamins and are also important in the making of blood pigments, bone, and hormones. The last of the seven components is water, which is used in most of the activities that take place in the chemistry of the body; it is also used in excretion and sweating.

VEGETARIAN DIETS

More and more people are now adopting a vegetarian diet. This can be a very healthy way of eating as long as care is taken to ensure good supplies of protein, plus essential vitamins and minerals, such as Vitamin B^{12} and iron. Vegetarians who eat eggs, fish, and milk should have few problems. Vegans do not eat any animal products and they must be particularly careful to enrich their diet with B^{12}. The great advantage of vegetarian food is that it reduces the amount of animal fats in the diet very considerably. Meat-eaters have to watch their diet much more carefully because the cholesterol in animal fat can cause heart disease.

WHY EXERCISE IS IMPORTANT

To be really fit we need both to maintain our strength by keeping our muscles active and to build up our stamina so that we do not tire quickly. It is also essential to keep our bodies agile so we can move joints freely. Ideally when we exercise, some of our sports and leisure time should be spent on building up strength, stamina, and agility. Swimming is the best all-around exercise there is.

INDEX

Figures in **bold** refer to captions.

ACKNOWLEDGMENTS

PHOTOGRAPHS

Biofoto Associates: 52 (left)
Bubbles: Jacqui Farrow 52 (right); John Garrett 44; Loisjoy Thurstun 9 (bottom), 19 (bottom), 35 (top); Ian West 32, 35 (bottom), 43 (top), 48; J. Woodcock 47 (bottom left)
The J. Allan Cash Photolibrary 61 (bottom), 69, 70 (bottom)
Camera Press John Doidge 17 (top left)
Bruce Coleman 75 (right); Patrick E. Baker 61 (top)
Lupe Cunha 56, 75 (left)
Sally and Richard Greenhill 21 (left)
Susan Griggs 30 (bottom); Michael St. Maur Scheil 9 (top left)
Octopus Publishing Group Ltd Sandra Lousada 36 (top), 59; Ron Sutherland 30 (centre)
Oxford Scientific Films London Scientific Films 11 (top); Herbert Schwind 47 (bottom right); D H Thompson 15 (top)
Science Photo Library CNRI 8 (top), 30 (top), 51, 62 (top and bottom), 65; DMIKRON 43 (bottom); Eric Grave 36 (bottom); Manfred Kage 13 (top right); NIBSC 57; David Parker 72; Petit Format/Nestle 67 (top and bottom); David Scharf 13 (top left); Secchi-Lecaque/Roussel-UCLAF/CNRI) 11 (bottom), 27; St Bartholomew's Hospital 55
Zefa 8 (bottom), 38, 41, 47 (top), 71 (bottom); Amthor 9 (top right); Norman 9 (centre)

ILLUSTRATIONS

Peter Bull 10-11, 12 (top), 13, 15 (inset), 16 (bottom), 17 (top), 26-27, 29, 31, 37 (bottom left and top), 42-43, 44-45, 48, 50-51, 52-53, 54-55 (top), 55 (right), 56-57, 58-59, 60-61, 62-63, 64-65
Garden Studio Darren Patterson 54 (bottom), 68-69, 70-71, 74
Frank Kennard 18-19, 20-21, 22-23, 40-41
Virgil Pomfret Agency John Bovosier 24-25, 33, 38-39, 46-47, 49, 66, 73; Bill Prosser 12 (bottom), 15 (bottom)
Tony Randell 14, 16 (top), 17 (bottom), 28, 34, 37 (bottom right)